AF862532

Back to the Ground

Olivier Labussière · Germain Meulemans ·
Céline Granjou · Adrien Baysse-Lainé ·
Pierre-Olivier Garcia
Editors

Back to the Ground

Knowledge, Politics and Practices of Remaking Earth Strata

palgrave
macmillan

Editors
Olivier Labussière
Pacte Social Sciences Research Center
Université Grenoble Alpes, CNRS
Science-Po Grenoble
Grenoble, France

Céline Granjou
National Research Institute
for Agriculture
Food and Environment
Grenoble, France

Pierre-Olivier Garcia
Pacte Social Sciences Research Center
Université Grenoble Alpes, CNRS
Science-Po Grenoble
Grenoble, France

Germain Meulemans
CNRS
Centre Alexandre-Koyré
Paris, France

Adrien Baysse-Lainé
Pacte Social Sciences Research Center
Université Grenoble Alpes, CNRS
Science-Po Grenoble
Grenoble, France

ISBN 978-3-031-88887-8 ISBN 978-3-031-88888-5 (eBook)
https://doi.org/10.1007/978-3-031-88888-5

© The Editor(s) (if applicable) and The Author(s), under exclusive license to Springer Nature Switzerland AG 2025

This work is subject to copyright. All rights are solely and exclusively licensed by the Publisher, whether the whole or part of the material is concerned, specifically the rights of translation, reprinting, reuse of illustrations, recitation, broadcasting, reproduction on microfilms or in any other physical way, and transmission or information storage and retrieval, electronic adaptation, computer software, or by similar or dissimilar methodology now known or hereafter developed.
The use of general descriptive names, registered names, trademarks, service marks, etc. in this publication does not imply, even in the absence of a specific statement, that such names are exempt from the relevant protective laws and regulations and therefore free for general use.
The publisher, the authors and the editors are safe to assume that the advice and information in this book are believed to be true and accurate at the date of publication. Neither the publisher nor the authors or the editors give a warranty, expressed or implied, with respect to the material contained herein or for any errors or omissions that may have been made. The publisher remains neutral with regard to jurisdictional claims in published maps and institutional affiliations.

Cover credit: Maram_shutterstock.com

This Palgrave Macmillan imprint is published by the registered company Springer Nature Switzerland AG
The registered company address is: Gewerbestrasse 11, 6330 Cham, Switzerland

If disposing of this product, please recycle the paper.

To the Glacier Blanc,
the groundwater of the Atacama Desert,
the granite bedrock of the Rhine Graben,
the former mining territory of Orbiel in the French Pyrenees,
and the phonolite domes of the Massif Central

Olivier, Germain, Céline, Adrien, and Pierre-Olivier

Preface

The approach proposed in this book emerged at the crossroads of various research projects led by Olivier Labussière and Alain Nadaï on geothermal politics, Céline Granjou on carbon sequestration by agricultural and forestry soils, Germain Meulemans on the emerging engineering of living soils, Adrien Baysse-Lainé on rural land dynamics under pressure from capitalist economies, and Pierre-Olivier Garcia on the political ecology of the Alpine glaciers. Each of us was looking at the ground from a different perspective—political geology, political ecology, science and technology studies, anthropology, geography—with the scientific fields involved spanning even broader areas such as geology, hydrology, ecology, glaciology, agronomy, soil ecology, and energy science.

Our discussions uncovered common ground, particularly concerning the roles of place and materiality, as well as the political and justice issues at the heart of the projects aimed at a green transition. These conversations fostered shared questions that extended beyond the often-assumed boundary between surface soils and rocky undergrounds. The processes we studied transcended this division and instead followed different forms and modes of geo-socio-technical "loops" and nodes along circuitous itineraries and transversal fluxes of water, bacteria, and contaminants.

In June 2021, we convened an interdisciplinary conference entitled "Soils and Undergrounds in the Socio-Ecological Transition." The conference had three objectives: first, to bring together researchers in the humanities and social sciences who focus on soils and subsoils; second, to examine how these elements are reinterpreted and repurposed through a series of promises and projects of green transitions; and third, to create transversal research pathways that link soils and subsoils while addressing the theoretical and empirical challenges they present in our disciplines.

In addition to the 45 academic presentations, we were pleased to include three keynote speakers: Nigel Clark (Lancaster University), Anna Krzywoszynska (University of Sheffield), and the "Collectif Glacier," an art–science group led by artist Olivier de Sépibus Thomaïdis. The scientific committee played an important role in reviewing proposals and chairing sessions: Xavier Arnaud de Sartre (CNRS, Pau), Soraya Boudia (Univ. Paris Cité), Sébastien Chailleux (Univ. de Pau et des Pays de l'Adour), Marie Forget (Univ. Savoie Mont Blanc), Sylvie Gentier (BRGM), Christelle Gramaglia (INRAE Montpellier), Brice Laurent (Ecole des Mines), Alain Nadaï (CNRS, Paris), Jérôme Poulenard (Univ. Savoie Mont Blanc), Thomas Shellenberger (Univ. de Haute-Alsace), and Olivier Vidal (CNRS, Grenoble).

Administrative managers from the various research units engaged in the conference were instrumental in making it possible. We extend our thanks for their unwavering support and skill in organizing meetings despite the challenges posed by the COVID-19 pandemic: Catalina Esparza (PACTE), Johanna Jamardo (PACTE), Arthur Larpent (PACTE), Nathalie Léardini (PACTE), and Véronique Strippoli (PACTE).

The conference received support from many sponsors, to whom we are once again grateful: the PACTE Social Sciences Research Center in Grenoble, the French National Research Institute for Agriculture, Food and Environment (INRAE), the University of Grenoble Alpes, the scientific network "Innovations and local transitions in mountain areas" (Labex ITTEM), the French Office for Geological and Mining Research (BRGM), the Alliance Ancre, and the French National Agency for Scientific Research (ANR).

A 2-day workshop was held in Grenoble, France, in November 2023 to discuss contributions and support the development of the collective volume. The POSCA (ANR-20-CE26-0016) and ComingGen (ANR-18-CE38-0007) initiatives backed this effort.

Finally, our best intentions to express ourselves in clear and simple English would have been futile without the meticulous linguistic and proofreading skills of André Crous, with whom we greatly enjoyed working. We sincerely thank the PACTE Research Center (Grenoble) and the Alexandre-Koyré Centre (Paris), which provided grants to access his expertise.

Grenoble, France	Olivier Labussière
Paris, France	Germain Meulemans
Grenoble, France	Céline Granjou
Grenoble, France	Adrien Baysse-Lainé
Grenoble, France	Pierre-Olivier Garcia

Contents

Notes on Contributors

Maxime Algis is currently completing a PhD in political sociology at LISIS (Paris, Gustave Eiffel University) an interdisciplinary research laboratory dedicated to the study of science and technology in societies. He holds a degree in social sciences from the Ecole des Hautes Études en Sciences Sociales (EHESS) and is both an urban planner and architect. His research focuses on the governance of degraded peri-urban soils in the Parisian metropolitan area and lies at the intersection of environmental sociology, political sociology of science and an analysis of land issues through the lens of the political sociology of economic activity. He recently published: Algis, M. (2024). Quelles trajectoires pour les sols dégradés périurbains? In R. Charvet, C. Mougin, & E. Rémy (Eds.), *Sols urbains, environnement et santé: repenser les usages* (pp. 109–125). Éditions Quæ.

Alexsandro Arbarotti is a sociologist who initially worked on water management and access conflicts in Brazil. He defended his thesis at the Federal University of São Carlos, entitled *Water Access Disputes and Hierarchies in Land Reform Settlements* (2018). He continued his research in France at the LEESU (Laboratoire Eau, Environnement, Systèmes Urbains, Ecole Nationale des Ponts et Chaussées), focusing on new urban practices linked to water management and sanitation, including transitional urbanism and community gardens. He later joined the Red Cross, using his skills to combat precariousness and poor housing.

Adrien Baysse-Lainé is a tenured researcher at the CNRS, the French National Centre for Scientific Research, and is affiliated with the PACTE—Social Sciences Research Centre at the Université Grenoble Alpes. He earned his PhD from the Université Lumière Lyon 2 in 2018. His research spans critical rural geography, legal geography, and social geography of the environment, with research interests including farmland concentration, land access regulation, forest struggles and the climatization of rural soils. He recently participated in a national expert study on soil quality indicators commissioned by four government agencies and ministries (Soil Quality—Toward an Indicator System for Public Policy). He serves on the editorial committees of the journals *L'Espace géographique* and *Géocarrefour* and is the treasurer of the French National Committee of Geography, a member of the International Geographical Union.

Julie Beauté is a PhD student in contemporary philosophy at ENS Ulm, PSL. Her research combines philosophical, biological and architectural analyses to explore new connections between architecture and ecology. Her dissertation focuses on ruins and carrion to highlight the human and non-human dynamics of architecture through the lens of decay. She draws on environmental humanities, new materialisms, feminist epistemologies and field philosophy. She is the author and co-author of several texts, including the "Feral" and "Microbes" entries in the *Handbook of the Anthropocene* (Springer, 2022).

Vincent Bos is a geographer specializing in the mining–energy nexus. His work focuses on the social and spatial dynamics of natural resource exploitation. He analyzes the reconfigurations of global production networks for lithium and solar energy and their associated materialities (infrastructures, logistics, etc.) in the context of the energy transition in Latin America (Argentina, Bolivia, Chile), Australia and Europe. In 2022, he joined the CEA's Institut de recherche et d'études en économie de l'énergie (I-Tésé) where he leads the "Pôle Ressources de la transition," which conducts prospective research on the evolution of supply and demand for critical minerals like lithium and cobalt.

Nigel Clark is a professor of Human Geography at Lancaster University, UK. He is the author of *Inhuman Nature* (2011), co-editor with Kathryn Yusoff of a special issue of *Theory, Culture & Society* on "Geosocial Formations and the Anthropocene" (2017) and co-author with Bronislaw

Szerszynski of *Planetary Social Thought: The Anthropocene Challenge to the Social Sciences* (2021). His current work is around connections between the inner and outer Earth, the evolution of human care, and deep histories of European violence.

Olivier de Sépibus Thomaïdis is a contemporary visual and photographic artist, as well as a beekeeper and a gardener. As a photographer, he initially worked for a climbing magazine before turning to photojournalism. Since 2004, his work has embraced a strong artistic dimension without abandoning its documentary aspect. He published a book titled *Retraite* (2004), featuring the "Chibanis" of the Belsunce district in Marseille, with poet Habib Tengour. Inspired by artists like Richard Long and Andy Goldsworthy, he develops his work in and with nature. For two decades, he has used the landscape as a lens to examine our sensitive connections with the living world, notably documenting the disappearing glaciers of the Alps across France, Italy and Austria and drawing our attention to the mountain as a subject of law.

Gaëlle Deletraz holds a PhD in geography and works as a research engineer at the Energy and Environmental Transitions research unit at the University of Pau, France. She coordinates the methodological components of projects within her unit and explores personal topics, such as the use of science fiction as a research tool on transition themes. She leads the PaSciFic program and has been actively involved in the non-profit sector for years, promoting knowledge on sustainability. Since 2017, she has collaborated with researchers to offer interactive conferences on science fiction and climate change at the intersection of her activities.

Sijia Du holds a master's degree from the University of Beijing and the Institute of Political Sciences, Paris. She completed an internship at the G-EAU laboratory in Montpellier in 2023.

Etienne Dufour has a PhD in urban and land-use planning from the Université Paris 1 Panthéon–Sorbonne and is the author of a thesis entitled "The end of recycling? Metabolic rift and biogeochemical policies in the Parisian region during the 20th century" supervised by Sabine Barles (2024). This work explores the long-term eviction of agricultural recycling practices for urban organic waste (household waste, material from cesspools and sewage) and focuses on the historical dynamics that resisted the destruction of organic matter. Since 2024, he has been carrying out postdoctoral research at the University of Reims under the supervision of Baptiste Monsaingeon, studying the history of plastic waste and the consequences of its proliferation.

Marie Forget has been an assistant professor in geography at the University de Savoie since 2012 and is a member of the EDYTEM (Environments, Dynamics, Mountain Territories) laboratory. Her research focuses on the use and preservation of resources, particularly the mine–water–energy nexus through the social construction of resources. She analyzes and studies territorial dynamics from local to global scales, with a focus on mountain territories in energy transitions. Her fields of study include the Andes and the Alps.

Pierre-Olivier Garcia is an associate professor at the University of Grenoble Alpes. He completed a Ph.D. from the same university in 2015. His research and teaching focus on land planning, the interplay between geographical environments and climate change, and the challenges posed by modern land planning. He examines how urban and touristic developments in high mountains and the politics of geodiversity conservation confront the non-modern impacts of natural phenomena (e.g. the melting of glaciers) and the emerging sensitivities tied to these experiences. He co-leads the master's degree program in "Geopoesice" (geography, power, spaces, justice, environments) at Grenoble.

Jérôme Gidoin has a PhD in social and cultural anthropology from Paris–Descartes Sorbonne. Since graduating in organic viticulture in 2018 from Amboise LPA, he has devoted part of his life to being a winemaker in the Loire Valley. In 2019, he founded *AnthropoVino*, an applied anthropology network serving stakeholders in the world of wine. He recently published "À la recherche d'une route des vins de la Loire à vélo" (2023) in the journal *POUR*.

Christelle Gramaglia is a sociologist of science and the environment at the G-EAU joint research unit (INRAE, Montpellier—France). She works on mining and industrial pollution, both on the production of metrological knowledge on risks in controversial situations and on ordinary, local experiences of environmental degradation. She is involved in several participatory and citizen science projects. She also teaches the sociology of science and technology at the University of Montpellier.

Céline Granjou is a research associate professor of sociology at the French National Research Institute for Environment, Food and Agriculture, affiliated with the University of Grenoble Alpes. She is the recipient of a national grant for the research project "The Promises of Soil Carbon Sequestration: Innovations, Organisations, Knowledge." Through this project, she investigates the evolving relationships with soils as carbon sinks and the politics of carbon removal including the challenges of carbon measurement and quantification. Her research also examines the increasing emphasis on climate issues within academic fields and disciplines. Her research interests span science and technology studies, environmental sociology and environmental humanities, with a focus on nature conservation, soil/society relations climate politics, the role of environmental knowledge, and future studies. She is an associate editor for the journal *Futures* and the *Revue d'anthropologie des connaissances* and has served as an expert assessor for the French National Agency for Research and the Belgian National Fund for Scientific Research and on more than 30 doctoral dissertation committees. She was an invited fellow at the University of Technology Sydney for 2 years from 2014 to 2015.

Maud Hetzel currently a doctoral research assistant in sociology at EHESS, holds degrees in social and economic sciences from EHESS and the Université Paris–Nanterre. Her research looks into the burgeoning realm of "green" businesses, with a particular focus on the food waste composting sector, examining the effective transformation of production methods and organizational structures.

Marc Higgin is an anthropologist and researcher at the AAU–CRESSON laboratory (UMR Ambiances, Architectures, Urbanités) at the Ecole Nationale Supérieure d'Architecture de Grenoble and an associate researcher at PACTE. He focuses on everyday practices of social life, particularly the role of the "non-human" in these ways of inhabiting the world. He engages in ongoing experimentation with different research practices—including different modalities of writing, film and the visual arts—to examine the sensuous, corporeal and affective dimensions of these experiences.

Mathilde Joncheray is an associate professor in geography at the Université Jean Jaurès in Toulouse, France. She has researched the role of mining in regional planning in the Congo–Brazzaville. A reader of science fiction, she has engaged with the PaSciFic project on the representation of cities and subsoils. Currently, she is working on the digital commons and the globalization of North–South exchanges.

Olivier Labussière is a tenured researcher in geography at the CNRS, the French National Centre for Scientific Research. He is also a member of the PACTE—Social Sciences Research Centre at the University of Grenoble Alpes, France. His work examines the relationships between energy, society and space, focusing on their emerging ecological issues. His research spans science and technology studies, geography and environmental humanities. He employs practice-based methods including documentary filmmaking to explore the dynamics of emerging energy landscapes and issues of justice. Currently, he co-leads the French National Observatory on Offshore Wind Power, which develops art–science collaborations and interdisciplinary research across social, ecological, and engineering sciences. This work addresses the social and ecological impacts of large-scale infrastructures. He also conducts research in political geology, with a recent study on geothermal energy in France.

Marine Legrand is an environmental anthropologist and researcher at the LEESU (Laboratoire Eau, Environnement et Systèmes Urbains) at the Ecole Nationale des Ponts et Chaussées. She earned her PhD from the French National Museum for Natural History in 2015 with a dissertation entitled *The Ecological Ordering of Urban Parks: Knowledge, Practices and Landscapes*. Her research in France involves ethnographic studies on the greening of land-use planning and building practices in industrialized contexts. Specifically, she focuses on the relationships between landscape design and biodiversity conservation, agri-urbanism, and ecological sanitation. She studies the interaction between informal uses and institutional approaches, infrastructures and living environments, as well as the flow of knowledge and imaginaries associated with emerging ecological practices. She strives to develop an open scientific approach oriented toward collaborative research-action, interdisciplinarity and the arts. Recently, she notably contributed to the book *With Microbes* (C. Brives, M. Rest, & S. Sariola, Eds.), published by Mattering Press in 2021.

Julien Merlin is a researcher in anthropology and sociology of science and technology, with a focus on activities related to the underground, such as mining, geothermal energy, and geology. He is particularly interested in how energy transition ambitions are reshaping our relationship with geological environments. He wrote his PhD thesis, "The mining trial: Indigenous people and the environment in the New Caledonian mining sector," at the Centre de Sociologie de l'Innovation (Mines ParisTech, PSL University). He subsequently completed postdocs on the "French mining revival" at the University of Lorraine, France, and on geothermal programs in France at the CNRS, PACTE—Social Sciences Research Center Grenoble.

Germain Meulemans conducts research at the intersection of environmental anthropology and science and technology studies. He is interested in recognizing soil and the multiple forms of life it harbors as social and political actors in their own right rather than simply as resources or developable land. His current research focuses on the role of urban soils in soil sciences, urban planning, and critical urban gardening. He also explores the development of experimental, collective ethnographies to address the politics of urban surfacing. He obtained his PhD in Anthropology from the University of Aberdeen and the University of Liège in 2017, working on Tim Ingold's Knowing from the Inside project. He has held postdoctoral positions in various STS (IFRIS, Sciences Po in Paris) and urban research (University of Grenoble) departments before joining the Alexandre-Koyré Centre (Paris) as a CNRS research fellow in 2021. He often works alongside artists, science historians, urbanists, and architects.

Delphine Montagne is a research engineer at the Energy and Environmental Transitions research unit at the University of Pau, France. She is a geographer and cartographer who supports researchers with projects involving maps and data. She has worked on the role of maps through works such as "Du réel aux virtuels, du virtuel aux réels : ce que la cartographie fait au monde" "Quand la cartographie devient bête" and "A la recherche des œufs de Pâques cartographiques"). Specializing in biodiversity, she has participated in several biodiversity projects, including the red list of spiders in France.

Alain Nadaï is a sociologist at CIRED, the International Research Centre on Environment and Development, which is part of the French CNRS. As an STS scholar, his research centers on renewable energy development and controversies, as well as the influence of numerical models on the construction of low-carbon futures. He has contributed as a leading author to the IPCC Special Report on Renewable Energy Sources and Climate Change Mitigation (SRREN).

Chloé Nicolas-Artero holds a PhD in geography and defended her doctoral thesis at the Centre de recherche et de documentation des Amériques (UMR 7227) at the Université Sorbonne–Nouvelle. Since 2016, she has taught geography at various universities. In 2021, she began a postdoctoral fellowship at the Center for Climate and Resilience Research in Santiago, Chile, and since 2023, she has been a postdoctoral researcher at Politecnico di Milano, Italy. Her research focuses on the social construction of water scarcities and their effects in semiarid regions. In 2024, she published the book *S'approprier l'eau: Droits, espaces et pouvoirs au Chili* with Presses Universitaires de Rennes.

Céline Pessis is an associate professor at AgroParisTech and conducts research in the environmental history of alternative agriculture in twentieth-century France. She explores the knowledge and technologies of agricultural modernization and how they are challenged by peasant movements, environmental limits and dissident scientists. Her work also illuminates various forms of socio-ecological conflict in France since 1945. She co-edited *Une autre histoire des 'Trente Glorieuses': Modernisation, contestations et pollutions dans la France d'après-guerre* (La Découverte, 2013) *Survivre et vivre: Critique de la science, naissance de l'écologie* (L'Echappée, 2014), and *Comment les machines ont pris la terre* (ENS Editions, forthcoming).

Mathilde Soyer is an environmental sociologist and an associate researcher at the LEESU (Laboratoire Eau, Environnement et Systèmes Urbains)—Ecole Nationale des Ponts et Chaussées, where she has led several projects related to water and sanitation policies. Her work focuses on innovation in stormwater management, the politicization of micropollutants in aquatic waters, and the construction of expertise in urban hydrology. Recently, she has been particularly involved in the deployment of dry toilets in various contexts, studying the motivations and obstacles at play, and how these devices are appropriated by managers and users.

List of Figures

List of Tables

CHAPTER 1

Exploring Geosocial Stratification in Times of Earthly Troubles

Olivier Labussière, Germain Meulemans, Céline Granjou, Adrien Baysse-Lainé, and Pierre-Olivier Garcia

List of abbreviations

IPCC	Intergovernmental Panel on Climate Change
BECCS	Bioenergy with carbon capture and storage
EGS	Enhanced geothermal system
HTG	High-temperature geothermal energy
HDR	Hot dry rock
STS	Science and technology studies

O. Labussière (✉) · A. Baysse-Lainé · P.-O. Garcia
Pacte Social Sciences Research Center, Université Grenoble Alpes, CNRS, Sciences Po Grenoble, Grenoble, France
e-mail: olivier.labussiere@umrpacte.fr

C. Granjou
National Research Institute for Agriculture, Food and Environment, Grenoble, France

G. Meulemans
CNRS, Centre Alexandre-Koyré, Paris, France

© The Author(s), under exclusive license to Springer Nature Switzerland AG 2025

O. Labussière et al. (eds.), *Back to the Ground*,
https://doi.org/10.1007/978-3-031-88888-5_1

1.1 Promises of Green Transition and the Reinvestment of Earth Strata

A wide range of climate-energy scenarios compete to define low-carbon futures, building on a variety of technological promises, such as renewable energies, nuclear power, and carbon capture and storage processes. Many of these scenarios target soils and undergrounds to mitigate climate change. The road to the green transition is paved with planetary, soil, and ground-based developments.

These technological promises are numerous, and their development relates to various societal and ecological contexts, scientific communities, and fields of expertise. They facilitate a broad range of processes for tapping into and reassembling Earth strata to address ecological and climate challenges.

For instance, geothermal energy leverages a diverse array of technological concepts and geological contexts to heat urban districts or generate electricity, practices that have been in place since the early twentieth century (Manzella et al., 2019). Ultimate nuclear waste is expected to be stored underground, following decades of being discharged into the sea (Hamblin, 2008). The extraction of energy and non-energy minerals has accelerated with the production of solar panels, wind turbines, and electric cars over the past few decades (Carrara et al., 2020; Vidal et al., 2013). Carbon capture and storage aims to bury carbon dioxide from industrial processes in the ocean depths or geological formations no longer used by the fossil industry, even spurring new energy industry developments (IPCC, 2005; Neri O'Neill & Nadaï, 2012; Chailleux, 2020; McLaughlin et al., 2023). Carbon mitigation and sequestration are just parts of emerging narratives that shape our understanding of earthly processes and our relationships with them. These narratives also encompass biofuel production, pollution remediation, green urban planning, and food security and water scarcity management (Gomiero, 2018).

These technological promises cannot be separated from ecological issues. The fragile living skin of the Earth, the "crucible of terrestrial life" (Hillel, 2007) at the interface of solid rocks and the atmosphere, has become increasingly central to discussions at the crossroads of politics, science, and land-use planning. The United Nations launched the "International Decade of Soils" in 2015 to rally public attention as experts warned that soil degradation rates were accelerating and threatening to surpass several of the Earth's boundaries (Steffen et al., 2015). More than

70% of the Earth's ice-free surface is directly impacted by human activities today, including infrastructure, cultivated lands, extensive grazing, and forest plantation. The intensified land use envisioned by some IPCC scenarios to develop BECCS (bioenergy with carbon capture and storage) and slow climate warming could increase the risk of famine due to land shortages,[1] accelerate habitat loss for living beings, and disturb the water cycle. As highlighted by the IPCC's (2019) special report, Climate Change and Land, the complex interplay between multiple fields of initiatives—energy, food, and biodiversity—raises new ethical and political questions. Several terrestrial ecosystems, such as agricultural and forest soils, wetlands, tidal areas, and oceanic algae, are recast as promising carbon sinks, provided critical issues of carbon enhancement and stability are resolved.

With the current hopes for soils and undergrounds to navigate the green transition, we are witnessing a broadening of interactions with the terrestrial environment. It is no longer viewed solely as a reservoir of resources—including minerals and fossil fuels—to be extracted. The aim is also to take advantage of the capacities and "services" offered by belowground spaces concerning the circulations of energy, water, and biochemical elements. One example is the development of "technologies of burying" in hopes of permanently disposing of unwanted waste and residues, ranging from domestic and nuclear waste to CO_2 surplus (Kearnes & Rickards, 2017).

These diversifying geo-socio-technical potentials (extracting, capturing, sequestering, and regenerating) are still partially described and analyzed, particularly in terms of the issues of knowledge and power associated with them. Several reasons can be identified for this analytical gap. For most of the twentieth century, the human and social sciences tended to view soils and the underground as resources for the development of human societies, focusing primarily on land as a surface atop which all social and biological existence unfolds, thus overshadowing the layers beneath. These perspectives rarely aligned with soil science's

[1] The special report *Climate Change and Land* (2019) underlines that 25–30% of the world's food production is either lost (in non-industrialized countries) or wasted (in industrialized countries), which suggests that a lot of land could be used for something else.

views on the dynamic nature of soils and their morphology but corresponded more with the relatively static agronomical and physicochemical approaches that dominated the latter half of the twentieth century.

In recent decades, however, the humanities and social sciences have developed concepts and methods in fields like political ecology, new materialism, environmental humanities, and science and technology studies that support a renewed ambition to tackle the complexity of socio-ecological interactions. This shift echoes a broader scientific effort by the humanities and social sciences to develop independently, in dialogue with other scientific communities (i.e., geology, ecology, soil sciences, agronomy), a conception of the Earth as a dynamic and intensive phenomenon, especially in the context of the Anthropocene debate. Science and technology studies, in particular, have started to understand soils and subsoils not as inert and passive substrates for human activities but as complex assemblages of biological and material entities with their own agency and relational power (Salazar et al., 2020; Granjou & Meulemans, 2023). The result is a diverse body of knowledge that varies in its theories, case studies, and societal interactions.

This research has largely developed within separate conversations regarding soils and undergrounds. Soils have been revisited in relation to theories of care (Puig de la Bellacasa, 2017), the microbial turn (Krzywoszynska, 2019), and citizen science approaches to soil pollution (Price et al., 2024). Meanwhile, undergrounds have been explored by researchers in political geography (Billé, 2020; Braun, 2000), political geology (Bobbette & Donovan, 2019), political ecology and geography of resources (Bebbington, 2012; Bridge, 2013; Bridge & Le Billon, 2013; Kama, 2020), and inhuman and decolonial geographies (Clark, 2011; Yusoff, 2024). These emerging fields of study cover various social aspects such as knowledge systems, power dynamics, and technical systems, which all contribute to our understanding of living soil and the underground. Engaging with them requires social scientists to grapple with technical issues and develop specialized expertise.

Beyond the social worlds that scrutinize them, their material qualities differ significantly. Neither soils nor undergrounds are static. Social analyses can focus on how they move and change in non-linear ways, how they shape and interact with human life and plans, and how they are entities with lives of their own. However, the types of life to be considered in each case are quite distinctive. Soils brim with teeming microbial and organic life, placing them at the core of terrestrial life-and-death cycles, while the processes of the mineral depths span geological eras. If their materiality is to be taken seriously, they might indeed require distinct social, cultural, and political theories.

But can these tracks remain separate when anthropogenic processes move layers and strata to the extent that it becomes extremely common for materials from deep rocky strata to surface, and vice versa? Can soils and subsoils be treated separately when fossilized life, trapped and transformed in the underground for millennia, is extracted, used, and burned as fuel on the surface of the Earth within a few years (Mitchell, 2013)? This process releases tons of carbon into the atmosphere, which recent political promises and initiatives now aim to recapture and store for the long term in agricultural fields and forest soils. How do we understand the new, unexpected assemblages and strata generated by the unearthing of mineral pollutants over decades of industrial mining exploitation, and the new, obdurate, and toxic associations and conglomerates they now form with soils, rain, aquifers, rivers, plants, and bodies? What happens when industry and modern capitalism accelerate or disrupt the flows and loops that connect soils and undergrounds in ways that dissolve any easy distinction between these categories? And how can theories about the Earth (i.e., the geosocial turn) and the living soil literature and soil humanities be brought together in this context?

1.2 The Need for a Collective Inquiry

In June 2021, the editors of this book organized a 2-day symposium to bring together social science research communities focused on soils and undergrounds. The aim was to showcase the richness of case studies and theoretical paths explored and to facilitate cross-disciplinary discussions.

It quickly became apparent that many participants were examining borderline cases where the distinction between soil and subsoil was not self-evident. Far from being taken for granted, this boundary was continually crossed and blurred by socio-material entities traveling along geological, geographical, and ecological vertical divisions and categories up to the surface, to watercourses, ecosystems, and the atmosphere. For example, work on mining waste depicted the trajectory of minerals extracted from the Earth's deep strata and deposited on the surface as residues that form a new stratum of anthropogenic soil. Authors studying atmospheric carbon storage projects—whether deep storage or sequestration in agricultural soils—highlighted the elementary circulations between geological oil deposits, atmospheric layers, and soils. Promises of carbon sequestration in agricultural soils also rely on a speculative equivalence between the geological processes of fossilized carbon formation in the long term and the rapid periods of carbon metabolism within fertile soils (Berta & Roux, 2024). These cases, which prompted empirical reflections on the material circulations between soil and underground, also required us to navigate between the theoretical repertoires of geosocial approaches and the soil humanities. The questions raised by these discussions gave rise to a collective inquiry that the editors developed further, involving two years of exchanges and a two-day workshop at the University of Grenoble in 2023.

The present volume is the result of these ongoing conversations. It addresses the joint evolution of soils and undergrounds in relation to various forms of "greening" such as green energy, metabolizing waste, green cities, and the broader, hoped-for "green transition" of the economy. It considers how a range of initiatives and experiments are fostering new collectives, practices, and sensitivities that offer a fresh perspective on the soil, land, and underground-based promises of more sustainable practices and development. These planetary promises may be more varied and complex than anticipated. Current technological developments engage multiple strata simultaneously and create a specific interplay of soil, land, and underground that sometimes generates unexpected ecological loops, cycles, and feedback with the atmosphere, the hydro-system, and the living realm. As such, remaking and reassembling Earth strata may reveal risky geo-socio-technical pathways, which do not necessarily result in reclaiming safe environments or achieving social-ecological healing. In some cases, they may lead to further risks (e.g., drought caused by hydrological imbalances, long-term pollution, microbial stimulation underground), managed by experimental geoengineering processes (seawater desalination station, excavation and containment of contaminated soils, and the use of bactericides to prevent the corrosion of wells, respectively) that disempower local populations in their relations with water, living soil, and the underground.

The book brings together empirical, critical, and speculative insights into the rise of various experiments, projects, and initiatives aiming to reconstitute, replenish, regenerate, and care for soils and undergrounds. We aim to explore how reassembling earthly strata contributes to producing new social organizations and socio-material relations with them. This collective investigation moves away from viewing them as objects of public policy, collective action, or instrumental projects to exploring emerging political spaces in the strongest sense, through which new collectives, practices (agricultural, energy, climatic, etc.), knowledge (geological, pedological, biological, etc.), and power relationships are organized.

To this end, the book experiments with thinking in terms of de-stratification and re-stratification issues and processes. What does it mean to live on a de-stratified Earth? How can we deal with what melts, flows, breaks away, accumulates, and contaminates? How can we understand re-stratification efforts (i.e., efforts aiming to regenerate, reshape, and replenish Earth strata), including citizen initiatives aiming to put carbon back into the ground and refertilize the soil, or the implementation of big industry-led and state-sponsored schemes to remediate soil pollution? These questions will be tackled through various methodologies, including ethnographic case studies, the history of geology and the soil sciences, the sociology of expertise, visual arts, and science fiction literature.

In the next three sections, we will present a review of the literature that distinguishes between three strands of work: geosocial approaches to the underground, soil new materialisms, and a relational approach to Earth strata. We will also explain how we found inspiration in them.

1.3 Geosocial Approaches to the Underground

Recent years have seen the emergence of a wide-ranging literature on the underground (Bebbington, 2012; Kinchy 2018; Marston & Himley, 2021; Woon & Dodds, 2021; Klinke, 2021; Bosworth, 2024).

Works in geopolitics have considered the underground in terms of knowledge, power, and property (Squire & Dodds, 2019). The "vertical" dimension of space has been studied in relation to historical connections between geological science and state power development (Braun, 2000), and more recently in relation to emerging volumetric issues of state sovereignty (Billé, 2020). Following this geosocial turn, the contributions to Bobbette and Donovan's volume, Political Geology (2019) address the role of geologists in shaping our understandings of Earth and society. This work focuses on "the Earth's organization into strata, and the depth of geological time and transformation" to explore the intersections of geology and politics.

In the field of energy resource geography, the literature emphasizes the knowledge, mediations, and rules that equip states and private groups with calculative capacities and control over materiality (Bridge, 2010, 2013; Huber, 2022). In the wake of the shale boom, some studies have addressed the scientific and political issues related to the assessment of unconventional gas resources in Europe (Kama, 2020; Kama & Kuchler, 2019; Kuchler, 2017). Beyond the diversity of their methodologies, this geopolitical literature highlights the use of calculative practices and instruments as some European countries suddenly benefited from large "volumes" of unconventional gas. These "volumes" have ignited a major controversy amid deep uncertainty over their local geological conditions. Thus, it is critical to carry out ethnographic inquiries into the construction of this unconventional "potential," while recognizing that this "potential" is neither predetermined by standard economic and technical production models nor solely the result of speculative practices in the face of uneven material events (Weszkalnys, 2015). In line with this, an ethnography of geologists, drillers, engineers, modelers, certifying officers, and marketers at work helps illuminate how gas potential is calculated and economically valued amid material, technical, and social uncertainties (Labussière, 2021).

Emerging writings in the social sciences also discuss the role of Earth's strata and promote a new "planetary thinking" (Clark & Szerszynski, 2021). Clark (2011) examined the cultural significance of the discovery of plate tectonics in the 1960s (1962–1968), suggesting that the awareness of geological dynamism has created a permanent state of bodily vulnerability for human life on Earth, in relation to earthquakes, volcanic events, and other violent upheavals of the Earth's crust. This realization has also fostered the development of new political organizations and philosophical ideas aimed at countering this vulnerability and instead highlighting human autonomy, freedom, and self-determination—echoing Enlightenment philosophical thinking. In this context, Clark develops decolonial approaches to soils and subsoils that challenge anthropocentric accounts of society and seek to decolonize social sciences from the belief in human exceptionalism inherited from Enlightenment philosophies.

Clark and other authors also suggest that the development of humankind has been closely embedded in the material capacities offered by soils and propose to delve deeper into unpacking the powers of soils—encompassing land and deep geological subsoils (Clark, 2011 and 2017; Clark & Szerszynski, 2021; Yusoff, 2018; see also Granjou & Salazar, 2019). They argue for the need to "geologize" accounts of social history crafted by human and social thinkers, to acknowledge that what is often understood as human innovations and historical creativity has been inspired, conditioned, and shaped by the material capacities and agencies of soils—including land cultivation, metallurgy, and fossil fuel combustion.

1.4 Perspectives from Soil New Materialisms

Social engagements with Earth's strata are not confined underground. Processes of deep exploration, large-scale extraction, or burying vividly illustrate the intensive connections between the geological and the living soils. This book aims to explore the circulation of matter between strata, following water, soil organisms, or plants, to contribute to producing critical knowledge about the geosocial assemblages that emerge from the green transition. This leads us to other strands of literature that have addressed the "Earth" not as the basal depths of geology but as the living and life-bearing substrates of surface soils, that is, the soil's new materialisms.

The politics of land and soil-related knowledge have fostered a substantial body of literature that unpacks the role of knowledge discourses and instruments in assembling land and soil resources and materialities. In a similar vein to the underground, this literature investigates how soil and land become objects of knowledge and management among various human groups and communities. Inspired by science and technology studies (STS), actor network theory, political ecology, and Foucauldian approaches to governmentality, authors posit that "soils" and subsoils, as entities, do not exist out there, ready to be surveyed and extracted as resources. Instead, they must be rendered visible and calculable through a range of theories, devices, and institutions. They examine the technologies of knowledge and prospecting (e.g., cartography, geological surveys, land assessments, and estimates) that contribute to their *construction* as intelligible and governable objects and resources (Tania Murray Li, 2014; Kon Kam King et al., 2018; Warkentin, 2006; Landa & Feller, 2009; Kon Kam King & Granjou 2020; Winiwarter, 2014; Duvall, 2011). In Li's account, devices such as fences, title deeds, and regulations change how qualities and values are attributed to land and soils, enabling them to participate in a form of social life organized around international trade rather than local dynamics.[2]

These accounts illustrate how values and properties are attached to various soils, thus creating conditions for the exploitation of soil resources and their circulation in markets simultaneously. They elucidate how processes of knowledge circulation and application contribute to the joint production of soils and society. Soils are no longer in the background, as humans integrate them into the social fabric. However, the vitality of soils is acknowledged only since humans themselves have devised ways to understand them as dynamic and lively entities. This approach, linked to Foucauldian social theory and political ecology, is crucial to us, as struggles over who lives on, uses, and defines the land lead to distinct socio-natural articulations and becomings. The depleted soil of capitalistic plantations differs markedly from the soil of the adjacent forested common land. Here, political and soil processes intertwine to (re)make strata.

Soil new materialism initially focused on living soils—the fertile layers of soils where plants grow—in an attempt to enrich the social sciences' understanding of our interconnectedness with belowground life and ecologies (Salazar et al., 2020). It addresses the practices and ethics of caring for the living soil and documents the emergence of new relations and alignments with soil biota, moving beyond the productivist and technoscientific practices of soil fertilization and control. Strongly inspired by new materialisms, multispecies ethnography, and theories of care focused on the living soil, the authors aim to advance the recognition of soil not merely as a stock or a set of inert resources for human use and exploitation. The work of philosopher María Puig de la Bellacasa (2014, 2015, 2019) represents an important source of inspiration here. She has spurred a focus on the active capacities of the diverse living things—such as earthworms, fungi, and bacteria—that inhabit, consume, digest, produce, and transform soils, as well as on the active and dynamic nature of (top)soil and its materiality as a heterogeneous mix of living

[2] In relation to the underground rather than land, Kama and Kuchler (2019) also explore "how the (sub)terrain is rendered as economically calculable and politically governable domain by virtue of different techniques and practices of geo-expertise" and aim "to account for the practices through which the subterranean becomes imagined and reappraised as stocks of resources that are deemed essential for future exploitation and consumption" (see also Kama, 2020).

organisms and dead or decaying matter, emphasizing the role of decomposition in soil formation and regeneration (Abrahamsson & Bertoni, 2014; Lyon, 2020). This scholarship is strongly committed to fostering more attentive and ethical relations with soil as a living, lively, and vulnerable ecosystem, deeply entwined with human life (Hird & Clark, 2013; Krzywoszynska, 2020). It accounts for the development of practices of soil attention, sensing, and care in both traditional agricultural settings (Lyon, 2020) and contemporary agriculture (Goulet, 2010), blurring the distinctions between biotic and abiotic, organic and mineral, living and inert, biological and geological, as well as natural and social (Meulemans 2019, 2020). Part of this scholarship diverges from new materialist authors, who often view soil mostly as a site of life's profusion, a hub of connections, symbiosis, and mutual flourishing between humans and soil biota. For instance, Ureta and Flores (2018) or Tironi (2020) highlight a different kind of soil agency, associated with the resistance of hard soils and the power of contamination from toxic soils—dust, leachates, residues—and their capacity to circulate in environments and bodies (Gramaglia, 2020; Hird & Clark, 2013). Focusing on the chain of events triggered by the unearthing of Earth layers through extractive mining activities, Gramaglia (2020 and this volume) investigates the continuous de-stratification process as toxicity spreads through various environments and bodies. She emphasizes the extraordinary resistance these new toxic assemblages and strata present to any effort of assessment, inventory, remediation and social-ecological healing. These authors remind us that soils are not always conducive to attachment and connection but also characterized by pollution and contamination and, more generally, not readily available for human relation and care. Soil humanities provide an important line of inspiration for us because they extend beyond geosocial literature in their deep interest in material processes and efforts to position the ground as an active participant in socio-environmental challenges and responses.

1.5 Giving Life to Strata: A Pioneering Relational Approach

In the past two decades, the scientific debate on the so-called "Anthropocene" has revealed and spurred a broad spectrum of research on the material and ecological participation of earthly processes in social life—a movement that some of the literature presented above is a part of. This led to questioning traditional distinctions like those between the living and the inanimate, the solid and the fluid, and the human and the non-human and opened up new avenues of research. While stratigraphers discussed how to produce stratigraphic evidence of the start of the Anthropocene in geological strata (Lewis & Maslin, 2015), humanities and social sciences debated the socio-material processes at work and the historical relations of power and exploitation associated with what stratigraphers term the Anthropocene (Bonneuil & Fressoz, 2013; Clark, 2013; Yusoff, 2018).

Before the specific debate on the Anthropocene, a constellation of philosophers had already accounted for the influence of Earth's dynamics on social life. For instance, Gaston Bachelard's advocacy for atomistic thinking invites us to embrace the diversity of the material states (e.g., fluid, soft, chunky, gaseous, etc.) that influence the changing world we inhabit (Bachelard, 2018[1933]). Other authors, such as Gilles Deleuze (1969) and Michel Serres (2000 [1977]), considered how a pre-Socratic view of nature might integrate the spiraling race of elements and social passions to depict the precarious trajectories of civilizations with the Earth. Deleuze and Guattari (1980) offered a pioneering relational framework to approach social engagements with a dynamic Earth, introducing the notion of a stratum and, by extension, thinking through processes of stratification and de-stratification. These philosophies of the Earth were an important source of inspiration for authors involved in the geopolitical and geosocial turn, such as Nigel Clark and Kathryn Yusoff.

In the text "The Geology of Morals (Who Does the Earth Think it is?)," Deleuze and Guattari (2005) caution that the Earth, as unstable, unformed matter, flows in all directions as free intensities. They argue that such vitality of terrestrial processes has not been acknowledged by modern thinkers, whose thinking tends to rely on static views of nature and, correlatively, stratified conceptions of the social. With the modern portrayal of a fixed Earth came the naturalization of these strata into resources readily available to social and economic exploitation.

Deleuze and Guattari conceptualize the Earth as a process of stratification, where strata are continually forming, merging states of nature and society. Thus, the creation of a stratum involves a dual operation—differentiating nature from society and recombining them in a normative sense of aligning nature with political and economic expectations. For instance, unstable terrestrial flows (e.g., the Earth's heat), often described and quantified by science and engineering as stable units (e.g., temperature gradients), exist independently of their relationships with aquifers, faults, or bacterial communities. This passivation of the material world facilitates its integration into functional social structures, driven by specific interests and temporalities. In this sense, a stratum is generative, producing both a state of nature (a workable geothermal underground) and a state of society (a geothermal economy). Following Deleuze and Guattari, a stratum does not always coincide with or depend on a specific Earth layer—as Kathryn Yusoff (2017) has pointedly illustrated with the notion of "geosocial strata."

Contrary to structuralist thought, Deleuze and Guattari argue that stratification does not equate to gaining control over nature. Instead, stratification and de-stratification operate concurrently: Being linked to earthly strata virtually connects the social to broader relational engagements with the Earth. The authors elaborate in four ways:

- The process of stratification may resemble parasitism—of Earth by humans—but it differs as both parts interact reciprocally: Each influences the other. This contrasts with the modern illusion of existing outside of nature, highlighting how humans are themselves stratified, dependent on the Earth's material life through culture, economy, and the technological mediations developed to inhabit the world.
- Thus, the Earth constantly interacts with the social through unpredictable geological and ecological cycles and connections across space and time. Similar to the relationship between the actual and the virtual, processes of stratification engage with a more complex earthly field—a surplus that extends beyond the immediate stratum we inhabit. These loops and interconnections, sometimes too diffuse or delayed to be directly observed, include phenomena like the gradual pollution of an aquifer or seismicity induced by a well.
- This means that, beyond the social realm, the ways humans deal with a stratum are diverse. Deleuze and Guattari observe that sensitivity to the Earth's variations is manifold. In addition to the use of geophysical instruments, this sensitivity may vary across cultural backgrounds. Some people can perceive things that others cannot; it is about engaging and experimenting with new capacities for sensing the Earth and working with its material and biological processes.
- Ultimately, the ways people are affected have a profound impact on their perception and their abilities to name, describe, understand the Earth, and articulate their relational experiences with earthly strata. Creating new languages (through stories, songs, poems) is a

powerful means of supporting social engagement with the land and the underground (see, for example, Keith Basso's study of Apache place names, (1996), or Edouard Glissant's writings about the Easter Island, 2007).

This Deleuzian perspective is further developed by Nigel Clark (2017) and Kathryn Yusoff (2017), who introduce the notion of "stratum" to distinguish it from the more conventional concept of territory. In territorial thinking, humans are seen as primary political agents in spatial transformation. In stratum thinking, the political concerns over space extend to terrestrial processes beyond (but still including) the Anthropocene. Stratum thinking does not overemphasize the ability of socio-political organizations to control earthly entities (e.g., tectonic plates, in the case of earthquakes), whose scale, temporality, and materiality transcend and resist such control. As defined by Clark (2013, 2017), a stratum is a geo-socio-political formation constructed by humans within the Earth's geological realm. A stratum is highly coded and embedded in historical power dynamics but remains open to the unbounded forces of the Earth. For instance, a plane dispersing chemicals over agricultural fields mobilizes various geological forces, including extracted chemical components and fossil fuels, to transform the land into fertile soil for a chosen plant species intended for international markets—thus demonstrating how industrial agriculture constitutes a stratum that integrates the Earth's layers and society through specific connections and configurations.

Such a relational approach introduces a stimulating way of discussing contemporary processes of de- and re-stratification. This can be illustrated by the recent debate between Bruno Latour and Frédéric Neyrat on living on a de-stratified Earth. Latour (2014, 2018) borrows the term "critical zones" from scientists who have developed a range of scientific observatories to monitor and measure the biogeochemical impacts of global environmental changes on various local sites and ecosystems. He redefines this concept to invite us to think in terms of "critical zones"—concrete local areas where life-sustaining conditions are threatened and diminishing, where socio-ecological struggles are emerging, and where new scientific assemblages are needed to envision other possible futures for the Earth and probe their political implications.

For Frédéric Neyrat (2018), the way critical zonists (primarily scientists) imagine developing more sustainable relations with Earth's strata—mostly through technical instrumentation, observation, and monitoring—is illusory, even dangerous, as it may reinforce the constructivist view of the planet as an exploitable resource. Referring to the "unconstructable part" of the Earth, he argues for a deep asymmetry between human and terrestrial processes. While Latour refers to "critical zones" as local sites where a shared cosmos is still to be created, Neyrat calls for a radical reform of our economies and lifestyles to experiment with alternative geosocial strata.

We propose that a relational approach to social engagements with Earth goes beyond the constructivist/anti-constructivist debate. By expanding the analysis beyond technoscientific processes alone, it is possible to gain, more nuanced view of contemporary processes of remaking Earth's strata. This perspective opens up the analysis to a diverse array of collective know-how, practices, and sensitivities that arise from transitory ways of living on and with a de-stratified Earth.

1.6 Bridging Earthly Conversations

Although rarely discussed together, the three approaches we outlined above can complement one another to help us understand contemporary reshaping of soils and subsoils. The post-Deleuzian approach to the subterranean shares with the Foucault-inspired literature on the politics of land and knowledge an interest in scrutinizing how scientific knowledge has contributed to framing Earth and soils as objects of knowledge and representation. It also shares with the literature on soil new materialism a commitment to paying closer attention to the materialities, agencies, and dynamics inherent to soils. However, these authors redefine the concept of soil agency, incorporating much broader scales and temporalities than those typically considered by soil new materialism: Soil agency is linked no longer solely to microbial life but also to the persistence of toxic dust and the movements and upheavals of the Earth's crust over geological timescales. All three approaches ultimately help us consider not only how various societies and peoples engage with soils and subsoils but also, symmetrically, how soils and subsoils contribute to shaping social communities and organizations.

Building on these authors' ongoing effort to decenter humans in the social sciences and humanities, this book seeks to address the co-constitution of society and (sub)soil materialities with a critical eye toward rethinking the role of the natural and material world in social and human sciences theories. At the same time, rather than simply adding perspectives, the book proposes to develop new transversal lines of research that connect hybrid processes and entities usually kept separated, beginning with lands, soils, and undergrounds.

In so doing, we aim to understand how soils and undergrounds come to be known, sensed, felt, assembled, managed, and inhabited—not only as presumed reservoirs of mineral and organic resources but also through various emerging practices, collectives, and sensitivities that aspire to regenerate and reconstitute Earth strata in a time of climate and ecological crisis. These emerging socialities, practices, and sensitivities contribute to the formation of new "Gaïan collectives" (Krzywoszynska, 2021) and "Earthly multitudes" (Clark & Szerszynski, 2021), whose analysis forms an original contribution of this book. The introduction sets the stage for a collective inquiry in which interdisciplinary and creative research play a leading role.

As a key figure in this relational approach, Nigel Clark's contribution has been positioned in the book's introductory section. His work has long advocated for an interdisciplinary approach to demonstrate how planetary thinking emerges from a dialogue between astrophysics, earth sciences, ecological sciences, and the humanities (Clark & Szerszynski, 2025). A nuanced understanding of Earth's material and ecological life sheds light on how humans develop their ways of living through specific associations with shifting terrestrial conditions. His chapter is especially thought-provoking, not only because of the questions it addresses—such as humanity's capacity to work with fire on our planet—but also because of the example of the value of a relational approach. In this chapter, the exploration of fire is not a moral issue, as is often the case in discussions of climate change linked to thermo-industrial civilization. Instead, Clark examines how multiple fire-centered practices interact with soils, living communities, rock layers, and, more indirectly, the seething, unlivable depths of the Earth's interior. Burning biomass, introducing it into rocky voids, and drilling into magma bodies can all be seen as ecologies of practice. As "fire knows no boundaries," it requires experience, knowledge, and care to maintain a stable relationship with it and mitigate potential planetary consequences. This chapter establishes the planetary scale as the setting for the collective inquiry that will unfold.

1.7 Organization of the Book

The book is organized into three sections that explore social engagements with a de-stratified Earth, processes of earthly re-stratifications, and emerging practices and sensitivities toward the Earth. Each of these lines of research invites an examination of the interactions between lands, soils, and undergrounds, tracing unexpected geo-socio-technical loops. The ambition is twofold: to test a relational approach to remaking Earth strata and to contribute to building an interdisciplinary research community.

Inhabiting a De-Stratified Earth

This first section addresses what it means to live on a de-stratified Earth. Drawing on both empirical case studies and artistic approaches, the chapters describe how unexpected material changes can be observed and made perceptible. The chapters illustrate how mediumslike photography, film, and intimate writing can reveal subtle, often imperceptible terrestrial transformations. This section also examines the emergence of alternative imaginaries of our connections to soils andthe underground. It offers a stimulating perspective on how we imagine living with Earth as an active agent and how its internal dynamics disrupt social and political orders.

Olivier de Sépibus Thomaïdis, a contemporary photographic artist, has spent over two decades using landscapes as a lens to examine our connections to the Alps. His work captures the disappearing glaciers across France, Italy, and Austria. This rapid transformation of the Alpine environment signals a crisis in how we perceive and describe these mountains. It is not just the mountains that are de-stratifying; our cultural connections to them are also unraveling. By photographing the mountains "at glacier height," he invites viewers into a landscape that challenges our acceptance, appreciation, and understanding. He encourages a shift from a contemplative to an investigative perspective, transforming the mountain into a tangible mystery that prompts us to reconsider our relationship with the land.

Gaëlle Deletraz, Mathilde Joncheray, and Delphine Montagne move beyond this photographic documentation of reality by studying a corpus of science fiction works from the 1940s to the 2020s. These books, comics, and videos depict characters and events set in post-apocalyptic worlds where the surface and the underground play opposing roles compared with the present: Humans are, in one way or another, forced to live underground, while the surface is perceived as a lost paradise. Their analysis outlines various forms of spatial and social organization in connection with subterranean environments.

Julie Beauté also draws on an artistic corpus, featuring paintings by Polish painter Jacek Yerka that bring spaces and strata into tension. Yerka's science fiction architecture challenges perspective norms and deconstructs the traditional vertical representation of Earth's layers, revealing the depths below the ground. His paintings offer more-than-human perspectives, presenting geo-narratives where the surface and the underground transcend their roles as passive receptacles.

Christelle Gramaglia and Sijia Du address the programs and initiatives to reclaim mining soils at a former industrial site in the French Pyrenees. They chronicle the successive remediation efforts developed by industrial and state-related stakeholders over 30 years and the gradual shift from extensive depollution efforts to more modest attempts to minimize risks posed by various toxic environments, such as soils, water, and plants. The authors analyze the sociotechnical imaginaries of the engineers, technicians, and experts involved in remediation programs, emphasizing their steadfast optimism and faith in scientific and technical capacities, including future innovations. This is despite the resistance and unexpected material qualities and movements of toxic pollutants that are unearthed and disseminated across various surface environments.

Earthly Re-Stratifications

Re-stratification technologies include a diverse array of practices, knowledge, and technologies aimed at reconstituting, reorganizing, and refilling Earth's strata on various scales—whether reintroducing carbon intothe soil, mitigating airborne toxins, replenishing fertile soil layers, or using controlled burns to blast away vast amounts of organic matter. This section examines these complex dynamics of de-stratification and re-stratification as a focus for investigation and analysis, probing what efficient, just, and inclusive re-stratification might entail. Subsequent chapters discuss the creativity and open-endedness of earthly re-stratification technologies through three case studies.

Maxime Algis analyzes the hesitance of different City of Paris departments to agree on a water management plan. It draws from participatory observations at inter-departmental meetings where the restoration of the natural water cycle—reinforcing rainwater infiltration by reopening urban soils—was discussed, along with its urban planning implications. The hypothesis posited is that divergent or even opposing conceptions of what constitutes "good" or "bad" soil are at the root of misunderstandings about how to green this sector of public action.

The case of lithium exploitation in Chile illustrates the complex interactions between the underground, the land, and the atmosphere, through what might be called a landscape of evaporation. The brines—a mix of water, salt, and lithium—have been socially constructed by the industry and the state as a mining resource rather than a water resource. This allows for its ongoing exploitation despite the disruption to the water cycle and notable socio-ecological impacts, such as water scarcity and the loss of wetlands in the Andes. The current necessity to transport and inject seawater to fully exploit the deposit follows the accelerated pumping and evaporation of subterranean aquifers. From a legal geography approach, Marie Forget, Vincent Bos, and Chloé Artero-Nicolas contribute insights on the legal dimensions of the social construction of nature, as well as the use of law in environmental conflicts.

Alain Nadaï, Julien Merlin, and Olivier Labussière explore the history of high-temperature geothermal energy (HTG), a pioneering geoengineering response to the 1970s oil crisis. The geotechnical concept involves tapping into the Earth's crust—the granite basement or bedrock—to harness its high heat. This chapter traces the evolution of this vision and the related scientific and technological concepts, including engineering projects like oil drilling, hydraulic fracturing practices, and nuclear waste burial, as well as scientific fields that were supposed to allow us to create these geothermal "loops." The analysis spans from the concept of hot dry rock (HDR), which originated in Los Alamos in the 1970s–1980s, to the later enhanced geothermal system (EGS) in Europe in the 1980s.

Emerging Practices and Sensitivities to the Earth

Since the beginning of the nineteenth century, industrialization, urbanization, and the rationalization of agriculture have created a "metabolic rift" (Foster, 2000) between areas of human settlement and the soils in their *peripheries. This section explores the efforts of citizen or professional collectives striving to reconnect with land and soils from a Latourian perspective of "reterrestrialization" (Latour,2018). These groups engage with the biological and geochemical cycles linking societies, cities, bodies, plants, and soil. The four contributions in this section examine collectives and practices of "caring for" soil (*Puig de la Bellacasa, 2017*) that not only advocate a discourse on soil and soil issues that is very different from that of mainstream agronomy but also seek to foster collective arrangements and lifestyles more attuned to the life and regenerative cycles of soils and ecosystems.*

Based on a study of documents (scientific and professional journals, institutional archives, etc.) relating to organic farming and agricultural modernization in the French context between the 1930s and the 1970s, Céline Pessis's chapter illustrates how the notion of "living soil" was historically constructed as both a scientific and practical category in the twentieth century. This notion, often presented as antagonistic to the utilitarian perspectives that dominated soil science and conventional agriculture, was redefined and sometimes downplayed by various networks of actors. However, Pessis reveals that the idea of living soil predates the alternative agricultural movements of the 1970s and was, in fact, a dominant theme within soil sciences in the first half of the twentieth century, central to their definitions of fertilization. This offers a more nuanced view of the living soil category between soil sciences and alternative agricultures, highlighting its shifting practical and political implications.

What practitioners define as soil life is also a key research question for anthropologists and sociologists working in today's society. Is it only the microbes, or do worms, moles, and plants too? How should one work with soil life? Independent anthropologist and winegrower Jérôme Gidoin explores the complex relationship between mechanical plowing, the use of chemicals, and the alliance with soil life in the practices of contemporary French winegrowers. In the realm of organic viticulture, the rejection of pesticides has led to increased use of plowing, which remains controversial: Does it inhibit or foster soil biodiversity? Gidoin's chapter shows that the simple dichotomy between intervention and laissez-faire, between working the soil and working *with* the soil, becomes blurred in light of differing definitions of what constitutes soil life, compelling the anthropologist to document a broad spectrum of approaches to defining and interacting with soil life.

The chapter by Marine Legrand, Etienne Dufour, Mathilde Soyer, Alexsandro Arbarotti, and Marc Higgin unpacks the historical use of human waste as fertilizers and investigates a variety of contemporary initiatives for implementing "dry" toilets in France, carried out by both institutions and NGOs. In an era dominated by chemical fertilizers in agriculture, they explore the various benefits of returning human waste to the ground and characterize the political and socio-economic critiques of agriculture and capitalism associated with these initiatives. They suggest that these initiatives demonstrate emerging concerns over soil fertility intertwined with power relations, economization processes, and ways of inhabiting the living world.

In line with recent literature on living soil, Maud Hetzel's chapter looks into the development of composting activities by documenting the rise of entrepreneurial businesses aimed at collecting and composting organic waste from residents and food retailers. Through an ethnographic inquiry conducted within recent composting companies and employing sociological concepts of entrepreneurship and the construction of causes, she documents the intersection of concerns over soil fertility and conservation with institutional issues of waste management. She illustrates how composting entrepreneurs, initially focused on food waste collection and management, become increasingly concerned with soil quality and fertility issues.

Competing Interests and Acknowledgments The authors have no conflicts of interest to declare that are relevant to the content of this chapter.

The authors thank the Social Sciences Research Center Pacte in Grenoble, the French National Institute for Research on Agriculture, Alimentation and Environment (INRAE), the University Grenoble Alpes, the Centre Koyré, the scientific network "Innovations and local transitions in mountain areas" (Labex ITTEM), the French Office for Geological and Mining Research (BRGM), the Alliance Ancre and the French National Agency for Scientific Research (project POSCA: ANR-20-CE26-0016, and ComingGen: ANR-18-CE38-0007).

References

Abrahamsson, S., & Bertoni, F. (2014). Compost politics: Experimenting with togetherness in vermicomposting. *Environmental Humanities, 4*, 125–148.

Bachelard, G. (2018 [1933]). *Atomistic intuitions: An essay on classification*. State University of New York.

Basso, K. (1996). *Wisdom sits in place: Landscape and language among the Western Apache*. University of New Mexico Press.

Bebbington, A. (2012). Underground political ecologies: The second annual lecture of the Cultural and Political Ecology Specialty Group of the Association of American Geographers. *Geoforum, 43*(6), 1152–1162.
Berta, N., & Roux, A. (2024). The endless expansion of carbon offsetting: Sequestration by agricultural soils in historical perspective. *Cambridge Journal of Economics, 48*(3), 451–470.
Billé, F. (2020). *Voluminous states: Sovereignty, materiality and the territorial imagination*. Duke University Press.
Bobbette, A., & Donovan, A. (2019). Political geology: An introduction. In A. Bobbette & A. Donovan (Eds.), *Political geology: Active stratigraphies and the making of life* (pp. 1–34). Palgrave.
Bonneuil, C. & Fressoz, J.-B. (2013). *L'évènement anthropocène*. Seuil.
Bosworth, K. (2024). Filling the hole? On new geographies of the subsurface. *Progress in Human Geography, 48*(2), 151–171.
Braun, B. (2000). Producing vertical territory: Geology and governmentality in late Victorian Canada. *Ecumene, 7*(1), 7–46.
Bridge, G. (2010). The hole world: Scales and spaces of extraction. *New Geographies, 2*, 43–49.
Bridge, G. (2013). Territory now in 3D! *Political Geography, 34*, 55–57.
Bridge, G., & Le Billon, P. (2013). *Oil*. Polity Press.
Carrara, S., Alves-Dias, P., Plazzotta, B., & Pavel, C. (2020). *Raw materials demand for wind and solar PV technologies in the transition towards a decarbonised energy system* (EUR 30095 EN). Publications Office of the European Union, Luxembourg.
Chailleux, S. (2020). Making the subsurface political: How enhanced oil recovery techniques reshaped the energy transition. *Environment and Planning C, 38*(4), 733–750.
Clark, N. (2011). *Inhuman nature: Sociable life on a dynamic planet*. Sage.
Clark, N. (2013). Geopolitics at the threshold. *Political Geography, 37*, 48–50.
Clark, N. (2017). Politics of strata. *Theory, Culture & Society, 34*(2–3), 211–231.
Clark, N. & Szerszynski, B. (2021). *Planetary social thought: The Anthropocene challenge to the social sciences*. Polity Press.
Clark, N., & Szerszynski, B. (2025). Interdisciplinarity and the earth sciences: Knowledge at the threshold. In A. Machin & M. Wissenburg (Eds.), *Handbook of environmental political theory in the Anthropocene*. Edward Elgar.
Deleuze, G. (1969). Lucrèce et le simulacre. In *Logique du sens*. Minuit.
Deleuze, G., & Guattari, F. (2005). *A thousand plateaus: Capitalism and schizophrenia* (Original work published 1980). University of Minnesota Press.
Duvall, C. (2011). Biocomplexity from the ground up: Vegetation patterns in a West African savanna landscape. *Annals of the Association of American Geographers, 101*(3), 497–522.
Foster, J. (2000). *Marx's ecology: Materialism and nature*. Monthly Review Press.

Glissant, E. (2007). *La terre magnétique: Les errances de Rapa Nui, l'île de Pâques*. Seuil.

Gomiero, T. (2018). Large-scale biofuels production: A possible threat to soil conservation and environmental services. *Applied Soil Ecology, 123*, 729–736.

Goulet, F. (2010). Nature et ré-enchantement du monde. In B. Hervieu, N. Mayer, P. Muller, F. Purseigle, & J. Rémy (Eds.), *Les mondes agricoles en politique* (pp. 51–71). Presses de Sciences Po.

Gramaglia, C. (2020). Saltkrake - Penser la « vitalité » des résidus miniers pour mieux appréhender leurs effets toxiques. *Revue d'Anthropologie des Connaissances, 14*(4).

Granjou, C., & Meulemans, G. (2023). Bringing soils to life in the human and social sciences. *Soil Security, 10*, Article 100082.

Granjou, C., & Salazar, J. (2019). The stuff of soil: Below-ground agency in the making of future climates. *Nature and Culture, 14*(1), 39–60.

Hamblin, J. D. (2008). *Poison in the well: Radioactive waste in the oceans at the dawn of the nuclear age*. Rutgers University Press.

Hillel, D. (2007). *Soil in the environment: Crucible of terrestrial life*. Academic Press Inc.

Hird, M., & Clark, N. (2013). Deep shit. *O-Zone: A Journal of Object-Oriented Studies, 1*(1), 44–52.

Huber, M. (2022). Theorizing the subterranean mode of production. *Political Geography, 94*, Article 102601.

IPCC. (2005). *Carbon dioxide capture and storage* (B. Metz, O. Davidson, H. de Coninck, M. Loos, & L. Meyer, Eds.). Cambridge University Press.

IPCC. (2019). *Climate change and land: An IPCC special report on climate change, desertification, land degradation, sustainable land management, food security, and greenhouse gas fluxes in terrestrial ecosystems* (P. R. Shukla et al., Eds.). Cambridge University Press.

Kama, K. (2020). Resource-making controversies: Knowledge, anticipatory politics and economization of unconventional fossil fuels. *Progress in Human Geography, 44*(2), 333–356.

Kama, K., & Kuchler, M. (2019). Geo-metrics and geo-politics: Controversies in estimating European shale gas resources. In A. Bobbette & A. Donovan (Eds.), *Political geology: Active stratigraphies and the making of life* (pp. 105–146). Palgrave.

Kearnes, M., & Lauren, R. (2017). Earthly graves for environmental futures: Techno-burial practices. *Futures, 92*, 48–58.

Kinchy, A., Phadke, R., & Smith, J. (2018). Engaging the underground: An STS field in formation. *Engaging Science, Technology, and Society, 4*, 22–42.

Klinke, I. (2021). On the history of a subterranean geopolitics. *Geoforum, 127*, 356–363.

Kon Kam King, J., & Granjou, C. (2020). Mapping soil, losing ground? Politics of soil mapping. In J. F. Salazar, C. Granjou, M. Kearnes, A. Krzywoszynska, & M. Tironi (Eds.), *Thinking with soils: Material politics and social theory* (pp. 39–54). Bloomsbury Academic.

Kon Kam King, J., Granjou, C., Fournil, J., & Cécillon, L. (2018). Soil sciences and the French 4 per 1000 initiative—The promises of underground carbon. *Energy Research & Social Science, 45*, 144–152

Krzywoszynska A. (2021). *Sensing soils to live with (in) Gaïa*. Keynote proposed for the conference *Soils and underground in promises of socio-ecological transitions*, Grenoble, 9–11 June 2021.

Krzywoszynska, A. (2019). Caring for soil life in the Anthropocene: The role of attentiveness in more-than-human ethics. *Transactions of the Institute of British Geographers, 44*(4), 661–675.

Krzywoszynska, A. (2020). Nonhuman labor and the making of resources: Making soils a resource through microbial labor. *Environmental Humanities, 12*(1), 227–249.

Kuchler, M. (2017). Post-conventional energy futures: Rendering Europe's shale gas resources governable. *Energy Research & Social Science, 31*, 32–40.

Labussière, O. (2021). A coalbed methane 'stratum' for a low-carbon future: A critical inquiry from the Lorraine Basin (France). *Political Geography*, "Earth Politics", *85*, 1023–1028.

Landa, E., & Feller, C. (2009). *Soil and culture*. Springer.

Latour, B. (2014). Some advantages of the notion of Critical Zone for geopolitics. *Procedia Earth Planetary Sciences, 10*, 3–6.

Latour, B. (2018). *Down to Earth: Politics in the new climatic regime*. John Wiley & Sons.

Lewis, S., & Maslin, M. (2015). Defining the Anthropocene. *Nature, 519*, 171–180.

Lyon, C. (2020). *Vital decomposition—Soil practitioners and life politics*. Duke University Press.

Manzella, A., Allansdottir, A., & Pellizzone, A. (Eds.) (2019). *Geothermal energy and society*. Springer.

Marston, A., & Himley, M. (2021). Earth politics: Territory and the subterranean–Introduction to the special issue. *Political Geography, 88*, Article 102407.

McLaughlin, H., Littlefield, A., Menefee, M., Kinzer, A., Hull, T., Sovacool, B., Bazilian, M., Kim, J., & Griffiths, S. (2023). Carbon capture utilization and storage in review: Sociotechnical implications for a carbon reliant world. *Renewable and Sustainable Energy Review, 177*, Article 113215.

Meulemans, G. (2019). Wormy collaborations in practices of soil construction. *Theory, Culture & Society, 37*(1), 93–112.

Meulemans, G. (2020). Reclaiming freak soils: From conquering to journeying with urban soils. In J. Salazar, M. Kearnes, C. Granjou, A. Krzywoszynska, & M. Tironi (Eds.), *Thinking with soils: Social theory and material politics* (pp. 157–174). Bloomsbury.

Mitchell, T. (2013). *Carbon democracy: Political power in the age of oil*. Verso.

Murray Li, T. (2014). What is land? Assembling a resource for global investment. *Transactions of the Institute of British Geographers, 39*(4), 589–602.

Neri O'Neill, R., & Nadaï, A. (2012). Risque et démonstration, la politique de capture et de stockage du dioxyde de carbone (CCS) dans l'Union européenne. *VertigO – la revue électronique en sciences de l'environnement, 12*(1). Retrieved from http://journals.openedition.org/vertigo/12172

Neyrat, F. (2018). *The unconstructable Earth: An ecology of separation*. Fordham University Press.

Price, H. L., Engel-Di Mauro, S., High, K., Ramírez-Andreotta, M. D., Ureta, S., Walls, D., Chie, W., & Kinchy, A. (2024). Soil publics: Regenerating relations with urban soils through citizen science. *Local Environment, 29*(4), 480–494.

Puig de la Bellacasa, M. (2014). Encountering bioinfrastructure: Ecological struggles and the sciences of soil. *Social Epistemology: A Journal of Knowledge, Culture and Society, 28*(1), 26–40.

Puig de la Bellacasa, M. (2015). Making time for soil: Technoscientific futurity and the pace of care. *Social Studies of Sciences, 45*(5), 738–748.

Puig de la Bellacasa, M. (2017). *Matters of care: Speculative ethics in more than human worlds*. University of Minnesota Press.

Puig de la Bellacasa, M. (2019). Re-animating soils: Transforming human–soil affections through science, culture and community. *Sociological Review, 67*(2), 391–407.

Salazar, J., Granjou C., Kearnes, M., Krzywoszynska, A., & Tironi M. (Eds.) (2020). *Thinking with soils–social theory and material politics*. Bloomsbury.

Serres, M. (2000 [1977]). *The birth of physics*. Clinamen Press

Squire, R., & Dodds, K. (2019). Introduction to the special issue: Subterranean geopolitics. *Geopolitics, 25*(1), 4–16.

Steffen, W., Richardson, K., Rockström, J., Cornell, S. E., Fetzer, I., Bennett, E. M., Biggs, R., Carpenter, S. R., de Vries, W., de Wit, C. A., Folke, C., Gerten, D., Heinke, J., Mace, G. M., Persson, L. M., Ramanathan, V., Reyers, B., & Sörlin, S. (2015). Planetary boundaries: Guiding human development on a changing planet. *Science, 347*(6223), Article 1259855.

Tironi, M. (2020). Soil refusal: Thinking earthly matters as radical alterity. In J. Salazar, M. Kearnes, C. Granjou, A. Krzywoszynska, & M. Tironi (Eds.), *Thinking with soils: Social theory and material politics* (pp. 175–190). Bloomsbury.

Ureta, S., & Flores, P. (2018). Don't wake up the dragon! Monstrous geontologies in a mining waste impoundment. *Environment Planning D, 36*(6), 1063–1080.

Vidal, O., Arndt, N., & Goffé, B. (2013). Metals for a low-carbon society. *Nature Geoscience, 6*, 894–896.

Warkentin, B. (2006). *Footprints in the soil: People and ideas in soil history*. Elsevier.

Weszkalnys, G. (2015). Geology, potentiality, speculation: On the indeterminacy of first oil. *Cultural Anthropology, 30*(4), 611–639.

Winiwarter, V. (2014). Environmental history of soils, In M. Agnoletti & S. Neri Serneri (Eds.), *The basic environmental history* (pp. 79–119). Springer.

Woon, C. Y., & Dodds, K. (2021). Subterranean geopolitics: Designing, digging, excavating and living. *Geoforum, 127*, 349–355.

Yusoff, K. (2017). Geosocial strata. *Theory, Culture & Society, 34*(2–3), 105–127.

Yusoff, K. (2018). *A billion black anthropocenes or none: A historically grounded and embodied understanding of geological transformation*. University of Minnesota Press.

Yusoff, K. (2024). *Geologic life*. Duke University Press.

CHAPTER 2

'Fire Knows No Boundaries': Ecologies of Fiery Practice on a Stratified Planet

Nigel Clark

2.1 Introduction: Fire on Earth

There's an eco-futurist saying that's been doing the rounds for over 50 years, and it goes like this: 'We are as gods, we might as well get good at it' (Stewart Brand cited in Brockman, 2009). This feels to me at once self-centered and short-sighted. The self-interested aspect is the most obvious. As a prognosis for the planetary predicament, this storyline affords responsibility to the same civilizations or social formations who got us into this mess. The restricted vision of the aphorism is more complicated. Paleoarcheological evidence—scant and provisional—points toward our distant ancestors, most likely *Homo erectus*, learning to handle fire a million or so years ago (Bowman et al. 2013). If this is the case, then the capacity to transform the entire known world—to set it ablaze from horizon to horizon—comes early in the human story. We might even say that as the first of the Earth's living beings to handle fire, the challenge for *Homo* was the long-term process of learning to temper these precocious

N. Clark (✉)
Lancaster University, Lancaster, UK
e-mail: n.clark2@lancaster.ac.uk

© The Author(s), under exclusive license to Springer Nature Switzerland AG 2025

O. Labussière et al. (eds.), *Back to the Ground*,
https://doi.org/10.1007/978-3-031-88888-5_2

'god-like' powers and become cautiously, caringly, human. What seems remarkable is just how well our genus and species managed this, right up until the last few recklessly incendiary centuries. Some of us still manage our fires exceedingly well.

Fire is one of many ways that humans have acquired geological agency over the millennia. But an ability to work with high heat has a special role when it comes to the way that the genus *Homo* has interacted with the layers or strata that help distinguish our planet. 'Fire,' muses philosopher Gaston Bachelard, 'links the small to the great, the hearth to the volcano, the life of a log to the life of a world' (1964: 16). Just as a concern with fire can offer ways to negotiate conceptually across spatial and temporal scales, so too does it draw us into more practical means by which localized activities come to matter beyond their specific site. In turn, attending to ecologies of fiery practice, in their manifold forms, points us toward skills, strategies, imaginaries that might be under threat and in need of protection or enhancement. It helps us identify ways of knowing and doing that are exposed to physical or epistemic violence, as well as those that incarnate and perpetuate such violations.

In other words, taking practices focused on fire as an entry point for thinking about the stratified layers of the solid Earth can help us decide what is in need of care. As social scientists Céline Granjou and Germain Meulemans (2022) make clear in their comprehensive review of a flourishing concern with soils in social thought, it is quite possible to conceive of the constitutive layering of the Earth as extending far beyond the human, as both preceding and exceeding human presence, while simultaneously exploring ethical questions arising out of our entanglement with soil ecologies. In a world where combustion too often seems to be dangerously out of control, however, we should not forget that some fires may also need care and sustenance.

In times of escalating climate anxiety, writes environmental historian Stephen Pyne, 'air has replaced humus as the ultimate yardstick of fire effects. The role soil served as an indicator of environmental health for agricultural societies, the atmosphere has assumed for industrial states' (1997: 320). But there are many kinds of fire, and as Pyne has long advocated, the alarming contribution that some forms of combustion make to atmospheric change should not prevent us from appreciating how carefully deployed flame helps ecosystems in many parts of the world. Neither should the realization there is a surfeit of a certain sort of anthropogenic combustion lead too quickly to the assumption that the only good fires are the ones that are immediately and intimately entangled with the living world.

Fire is the visible effect of combustion—the ignition-triggered process in which energy held in the atomic bonds of a fuel is rapidly released through oxidation (a reaction with oxygen or an oxygen-rich compound) generating heat and forming new chemical bonds (Smil 2006: 10). It is also an exothermic process: a chain reaction in which the heat produced ignites still more fuel, which helps explain why fire is both an efficacious and a risky tool for managing living landscapes (Clark 2023). As Pyne (1997: 3) reminds us, Earth is the sole astronomical body in our solar system where fire is present—our home planet being the only place where there is an oxygen-rich atmosphere, fuel stocks, and sources of ignition. Even then, it took a planet-wide oxidation event, the rise of multicellular organisms, and the colonization of land by tough-bodied plants for fire's three ingredients to fuse into a workable assemblage—with the first evidence of charred biomass showing up in the Silurian Period some 420 million years ago (Bowman et al. 2013).

Besides the cosmologically rare phenomenon of fire, there are more common ways that heat—energy transferred from one system or region to another as a result of temperature difference—contributes to the formation and transformation of planets and other astronomical bodies. Though they are defined in various ways, planets can be characterized as roughly spherical bodies of condensed matter that occupy a thermal mid-range between the cold of space and the intense heat of stars. Additionally, as a result of gravitational forces acting on their constitutive chemical elements, planets organize themselves into concentric compartments or spheres (Dick 2019: xxxiv-v). For many planets, however, this is not the end of the story. Flows of energy from a parent star (if it remains part of a solar system) and from its own interior can prevent a planet from descending into equilibrium—leaving its compartmentalized layers crosscut by gradients of temperature and pressure. In this way, most planets remain open to thermally-driven evolution in which their subcomponents continue to transform, interact, and even reorganize themselves into new configurations (Clark and Szerszynski 2021: 79–80).

One way of conceiving of the Earth, then, is as a particularly dynamic version of a concentrically layered, thermally evolving planetary body. Such a view, I suggest, can broaden our perspective on the way different human collectives use high heat, helping us to situate human fire-wielding practices not only in particular places on the globe but in relation to the Earth's ongoing thermal evolution. In the words of North American Indigenous ecologist and cultural burning practitioner Frank Kanawha Lake: 'fire knows no boundaries, and it will move right across them' (Lake and Nelson 2022): an idea I want to extrapolate upon to show how a range of skilled human collectives have utilized high heat to negotiate junctures or thresholds in the body of the Earth.

After mapping out some of the different modes of conceptualizing human relationships with a dynamic and stratified planet, I look at how specialized fire-focused social practices have engaged with three distinct planetary zones or layers: the near-surface ecosystems of soil, the subterranean lithic strata that compose the Earth's crust, and the even deeper formations and forces of the inner Earth. Tapping into and channeling the matter energy of each of these structural components of the Earth, I argue, offers certain affordances for human agents—but also comes with its own risks. This raises questions not only about how practitioners acquire expertise but how they deal with the inescapable uncertainty that attends any intervention in the forces and dynamics of our planet.

2.2 Social Engagement with a Stratified Earth

The late 1960s–early 1970s was a turbulent but generative moment for thinking about the Earth, with breakthroughs in theorizing plate tectonics, extraterrestrial impacts, the coupling between living and nonliving components of the Earth, and the alternation of gradual and rapid evolution (Brooke 2014: 25–36). At first largely impervious to these developments, critical social thought's growing suspicion of abstract, impartial, and universalizing truth claims would later prove less than hospitable to planet-focused scientific inquiry. With the rise of Anthropocene hypothesis, disquiet over the issue of 'who speaks for the Earth'—whose perspective on the planet comes to matter and whose visions of the Earth and cosmos are precluded —have only intensified (Lövbrand et al. 2015).

There's a risk, in this context, that social science and humanities scholarship slips into a binarized thinking: a dualism that opposes the geologic, the planetary, the deep temporal to the lived, the local, the everyday. But recent geoscience encourages us to view the Earth in terms of its capacity to self-organize into a novel operating system, and as Bronislaw Szerszynski and I have sought to show, this idea of a 'planetary multiplicity' can enrich rather than detract from the interest in the difference or diversity that is so much a part of our humanness (2021: 8–9, 88–90). Alongside questions about how we are positioned in social worlds, such an approach prompts us to consider where and when we are located in the ongoing self-differentiation of the Earth—and to ask what planetary elements, powers, and processes human agents have joined forces with on the way to becoming who we are.

As the human counterpart to planetary multiplicity, we have couched the term 'earthly multitudes' (Clark and Szerszynski 2021: 9–11, 54). Earthly multitudes are social collectives who both convene around particular resources generated by past geological transformations and intervene in ongoing dynamic planetary processes (see also Labussière 2021). To be part of an earthly multitude, as we all are in more and less direct ways, is to share a particular orientation toward the affordances of the Earth, to at once shape and be shaped by the multiplicity that inheres in Earth systems at every scale from the local to the planetary. As such, an earthly multitude has a lot in common with an 'ecology of practices': Isabelle Stengers's term for ways of thinking about and engaging with the practical problems of worldly existence (Stengers 2005). Although Szerszynski and I are less willing than Stengers to define our 'multitudes' in opposition to dominant earth-oriented groupings or modes, we share with her a sense that the practices and practitioners in question do not stand outside the natural world but rather operate in a state of ongoing, open-ended entanglement with the matters of their concern.

There is also significant overlap between our earthly multitudes and the skilled actors whose conceptual and practical engagement with the critical zone was of such interest to Bruno Latour in his late work. The critical zone is an interdisciplinary scientific term for the life-suffused section of the Earth that reaches from the lower atmosphere to the bedrock: a slender but dynamic envelope, as Latour and his colleagues noted, which presents challenges requiring the collaboration of 'hydrologists, soil scientists, geochemists, geomorphologists and geophysicists, and ecologists' (Arènes et al. 2018: 121). While our earthy multitudes are a more inclusive array of agents than Latour's 'critical zonists,' what they have in common is an immersive, experimental orientation to a specific zone or region of the Earth.

Above all, if we are working with a western inheritance, any critical or speculative account of human engagement with self-organizing and stratified Earth must pay its dues to the 'stratoanalysis' of Gilles Deleuze and Félix Guattari (1987: 43). While not strictly reliant on a geological model, their conceptualization of a material-energetic world composed both of relatively consistent, tightly bound 'strata' and more free-flowing 'deterritorialized' elements clearly took inspiration from the Earth sciences. As with Stengers's ecologies of practices, Szerszynski and my concept of earthly multitudes owes much to Deleuze and Guattari's avowal of an experimentalism geared toward concrete structural components of the Earth—each with its own properties and potentials. 'Lodge yourself on a stratum,' they advised: 'experiment with the opportunities it offers, find an advantageous place on it, find potential movements of deterritorialization ... experience them, produce flow conjunctions here and there' (1987: 161).

Experiments, of course, don't always turn out as expected. And this has ramifications. To be operating in the thick of things, to intervene in a dynamic world, as Stengers would have it, is to be implicated and beholden. 'Practitioners have obligations,' she insists. 'This is the primordial fact for an ecology of practice' (2005: 190). Likewise, what attracts Latour to critical zoners and others who intercede in complex, troublesome fields, is the way they are confronted by their own 'terrestrial entanglement'—which can help them to see themselves not simply as observers but as co-composers of a common world (Arènes et al. 2018: 122; Latour, 2014). Or as science and technology studies scholar María Puig de la Bellacasa elaborates with particular attention to the human relationships with soils, a sense of pervasive, inextricable co-dependence draws us from matters of concern to matters of care (2017: 30–48).

Deleuze and Guattari too like to picture us proceeding from the midst of things. But the significance of strata in their thinking also pulls in other directions, for it attests that our own spheres of activity are underpinned by layers of existence that are vast, deep, and forceful. When human practitioners of any kind tap into the materials of another stratum—be it organic, geological or cosmic—they inevitably imbibe something of a world that radically exceeds their own lived experience or imaginings. Consequently, Deleuze and Guattari's advice to experiment with care is always in part a reminder that the forces which bring novelty and change into our lives can also overwhelm us (1987: 502–03).

Taken together, these various conceptual approaches help us to see that the layers or components of the spherical Earth have their own properties, dynamics, and forces—which means that engaging with them calls for specific skill sets, distinct experimental modes, and different kinds of care.

This would hardly come as a surprise to most of the earth-oriented practitioners or multitudes who have been intervening in planetary processes in various ways for countless generations. And those who work closely with fire to access and modify earthly matter, I argue, tend to be especially attuned to such particularities. Viewed through a wide-angle lens, the question of how humans have joined forces with the soil, the stratified lithic crust, and the inner Earth draws us into conversation with richly differentiated ecologies of fiery practice—or fire-using multitudes. While such collective endeavors often have just, generative, or sustainable aspects, it is important to note that some practices or forms of expertise may be careless, exploitative or unjust—though making such judgments its itself far from straightforward (Clark and Szerszynski 2021: 97).

2.3 Fire and Soil

Through deepening conversation with natural scientists and soil-oriented practitioners, some social thinkers are coming around to viewing soils as seething, dynamic, but also vulnerable assemblages of living creatures, organic and inorganic matter (Lyons 2020: 170; Puig de la Bellacasa 2017: 191). Soil, they point out, is where the solar-powered biosphere, hydrosphere, and atmosphere meet and mingle with the rock that comprises the Earth's crust. Here, in what is little more than a film or glaze at the planetary scale, a teeming array of micro and macroscopic organisms compose a world for themselves out of water, gases, particles of weathered rock, their own dead bodies, and energy-rich-tissues from photosynthesizing plants.

Hans Jenny's canonical model of soil formation from the 1940s recognized five universal factors: climate, organisms, topography, parent material, and time (Lyons 2020: 46), a schema to which humans were subsequently added. But as wildfire researchers Cristina Santín and Stefan Doerr add, 'fire is currently regarded by some as the seventh soil-forming factor' (2016: 1). While decomposition—breakdown of organic matter by microbes and larger invertebrate soil fauna—is crucial to the formation and development of soils, fire-focused researchers remind us that fire also releases nutrients and carbon by unraveling what photosynthesizing plants have assembled (Pyne 2001: 8).

If fire has yet to feature prominently in the current embrace of soil by social scientists, we should note that soil scientists are candid about their own shortfalls in understanding impacts of fire on soils and their constitutive biotic communities. As Yamina Pressler and her colleagues conclude in a comprehensive review:

> We currently do not have sufficient data to determine the consequences of shifting fire frequency, severity, seasonality, size and duration on the structure and function of the entire belowground community or the physical, chemical and biological interactions that give rise to these responses (Pressler et al 2019: 321).

Fire's late arrival in soil science needs to be viewed in the context of an earlier and generalized repudiation of broadcast burning emanating from Europe. As Pyne recounts, regular burning of fallow and periodic firing of forest was a cornerstone of European agroecosystems for millennia, while in many parts of the world—especially the tropics—fire has been an essential preparatory phase of both shifting and sedentary agriculture (2001: 88). But 'enlightened' agronomists in a modernizing Europe became convinced that surplus organic matter needed to be cycled back into soil or forest rather than 'squandered' in flame. With European global expansion, this geographically and culturally specific renunciation of fire was fashioned into a set of prohibitions that cultivators, pastoralists, and forest dwellers across the world were compelled to abide by, whatever their traditions and their local fire regimes (Clark, 2018).

As we might expect, there was widespread resistance, as customary fire practitioners reasserted their right to apply flame when and where they felt their lands needed it. As many Indigenous or traditional fire experts anticipated, indiscriminate fire suppression has had catastrophic consequences for foraging, cultivation, and other aspects of local culture, for habitat and biological diversity, and for wildfire risk (Kimmerer and Lake 2001, Pyne, 1997, 303–8, 325*)*. In particular, escalating wildfire hazards have prompted many western land managers in fire-prone areas to reconsider controlled burning. So too has there been growing appreciation of the way that traditional cultivators—in the Amazon, northwest Africa, and elsewhere—have generated exceptionally rich soils through protracted application of charred organic matter. While some commentators latched onto the potential of 'pyrogenic' carbon-enriched soils to function both as carbon sinks and as a means of enhancing agricultural productivity, critical social scientists have responded by noting the inflated promises and the real possibility that land grabs for large-scale carbon offsetting could undermine the very practice of local, site-specific soil enrichment that inspired the turn to 'dark earths' (Leach et al. 2012).

In a more generalized reappraisal of the impacts of fire on soils, Santín and Doerr (2016) distinguish between the deleterious impact of large, high-intensity fires on soil formation and organic communities and the more positive effects of frequent lower intensity fires, including those set by skilled practitioners. While stressing the importance of local variability of the factors that condition soil-fire interactions, they willingly conclude that 'fire can increase soil fertility, organic carbon content, weathering and, ultimately, soil formation, particularly in areas where limited topography or rapid vegetation cover limit post-fire erosion' (Santín and Doerr 2016: 4). Both attunement to the effects of repeated fires and sensitivity to the fine-grained, heterogeneity of the fire-life-soil nexus, it should be added, are staples of traditional land management (Kimmerer and Lake 2001). In the words of Frank Kanawha Lake: '(t)ribes bring a traditional ecological knowledge of plants, animals, fungi, tribal subsistence and burning practices' (cited in USDA, 2019). On another continent, Indigenous Australian ecologist and member of the Tagalaka people, Victor Steffensen describes how different trees are associated with a range of soils requiring specific kinds of fire, and speaks of the importance of timing of fires so that accumulated grass and leaf litter does not fuel a fire that is too hot for soil to bear (2020: 41–42, 164–5). As he passes on the advice of an elder: 'you have to have moisture in the soil to at least half a foot to a foot deep. Then you can burn ... and the vegetation will come back quicker and thicker' (2020: 45).

With their stress upon entangled organic and inorganic processes, Indigenous affirmations of a fire-enabled caring for country have much in common with soil-oriented social science. Much of this recent 'soiling of the social' seeks to counter soil-exhausting agriculture with agroecological practices that will help sustain soils and their constitutive living communities—professing what Puig de la Bellacasa describes as 'an ethos that contributes to the cocreation of a particular ecology and the mutual multilateral obligations and interdependent doings it entails' (2017: 201). It's worth recalling, however, that the deep-seated disavowal of fire in modern western thought and practice was propelled by the sense of its unboundedness, its exorbitance, its raging afront to 'restricted' economies centered on closed circuits. In contrast, there is something about the metamorphic power of fire in Indigenous ecologies of practice that seems at once to intensify and gesture beyond the tight loops of reciprocity evoked in much recent by western ecologically informed social thought. This opening outward– the excess we can detect in Lake's assertion that 'fire knows no boundaries'—has implications for the way care is construed. While much traditional land care entails relatively safe and contained burns, there is nonetheless an awareness that fire can escape, that fire regimes will shift, a sense that lives have been put on the line and will be again. Or as Steffensen puts it, reflecting upon the coming of climate change after centuries of interrupted cultural burning, 'it is better late than never; even if it was too late, then we must at least die trying' (2020: 211).

Unsurprisingly, when the stakes are this high, the Indigenous fire-life-soil nexus is frequently presided over by ancestors, spirits, and other figures who help mediate between the past, present, and future, or between this world and other worlds. When Kimmerer and Lake speak of 'a spiritual responsibility to carefully use fire to multiply life' (2001: 40), we might read this as evoking a multiplicity that inheres in the Earth and cosmos itself—in a way that recalls Deleuze and Guattari's cautioning about joining forces with potentially overwhelming powers. But so too should we recognize that fire's capacity to move across boundaries and junctures can reach deeper than the soil. If Steffensen's observation that '[f]ire is a commitment that goes on forever into the future' (2020: 87) speaks to fire-focused multitudes to come, so too from another angle does it gesture toward the deep time of the geologic.

2.4 Fire and Lithic Strata

Examining a sedimentary sequence north of Sydney, paleoecologists Manu Black and Scott Mooney found a spike in charcoal deposits—indicative of significant increase in the prevalence and intensity of fire in southeastern Australia around 5700 years ago (2007: 47). Triangulated with contemporaneous evidence from the Pacific Basin, Black and Mooney interpret this wildfire irruption as the signature of a threshold transition in the planet-girdling ocean-atmosphere system known as the El Niño Southern Oscillation. In the Australian charcoal record, the fierce new fire regime looks to have been sustained for the next 2500 years. Subsequently, there is a notable shift toward more frequent but much lower intensity fires, without evidence of further climate change. Black and Mooney's tentative explanation is that by this time, Aboriginal land managers had adapted their pre-existing cultural burning strategies to such a degree that they were able to successfully mitigate the intensities of El Niño's wet–dry fluctuations and to pre-empt the risk of vast and deadly conflagrations (2007: 50).

In this way, much as the Anthropocene hypothesis proposes that accelerating fossil fuel combustion over the last century will leave its trace in the geological strata now in formation, Black and Mooney's data suggest that, deep in the Holocene, Indigenous fire-wielding multitudes were already leaving indelible impressions in the regional sedimentary sequence. And just as fire has assisted human agents in probing the possibilities of the life-infused outermost envelope of the Earth, so too has the light and heat released by combustion enabled a diurnal, surface-dwelling creature to venture into pre-existing subterranean spaces—and eventually to carve out our own geological fissures and voids (Clark, 2021).

As Pyne (1994) contends, the capture of fire by a living creature marks a turning point not only in the evolution of our own genus but in the history of our planet. By the same logic, we can consider the transfer of flame from its essentially surficial niche into the lithic strata as another significant juncture for both us and the Earth. One early manifestation of this vertical enfolding of fire is the human inhabitation of caves. Though the vernacular term 'cave man' may give a misleading impression of the amount of time mobile, foraging Paleolithic people actually spent in geological voids (Isabella 2013), the artwork that adorns hundreds of excavated caves suggests we shouldn't underestimate the cultural significance of the firelit move into a stony subterranean milieux.

While their original meaning remains enigmatic to modern observers, the multitudes of acutely observed animal figures that play across rock ceilings and walls are indicative of the symbolic power emanating from the introduction of flame into subterranean space; an observation lent weight by speculation that the flickering light emitted by grease lamps would have helped mobilize or 'animate' these painted forms for their intended audiences (Zorich, 2014). Key pigments used in Paleolithic art, such as ochre, manganese oxides, hematite, and calcite were themselves sourced from caves, and one of the oldest known continuous mining operations—Wilgie Mie in the Weld Ranges of Western Australia—is a source of ochre for the Wajarri Yamatji people. Over some 27,000–40,000 years, Aboriginal miners have extracted an estimated 40,000 tons of rock bearing iron-rich red and yellow ochre that has been used for ceremonial, aesthetic, and medicinal purposes across much of the Australian continent (Australian Government, 2011; Hunt 2019: 144–180).

The miners of Wilgie Mie have used stone hammers and fire-hardened wooden wedges to crack open rock, while other Aboriginal artisans have long used high heat to render stone more workable and to change its color. Indeed, across many regions, people long familiar with the power of fire to alter the properties of wood, flesh, and bone also probed its potential to modify other elements. Perhaps drawing upon long-term observation of the impact of hearth fires on surrounding earth, ancient craftspeople learned that high heat, carefully applied, could harden soggy clay into a robust, impermeable, solid. If not at the planetary scale that the Anthropocene hypothesis proposes, we can view this power to transform the structure of inorganic matter as a mode of human geological agency, and as the crux of a new kind of fire-focused earthly multitude. As geologist Simon Wellings (2016) explains, high temperatures and the pressure of rock layers stacked above will eventually transform sedimented silt, mud, and clay into metamorphic rock. 'Taking clay and baking it in a kiln is the same process—a human-created form of metamorphism.'

The related heat-induced transmutation of crumbly ores into lustrous, ductile, and durable metals took the social involvement in the deformation and reformation of layered rock fabrics to another level. With the rise of metallurgy in the ancient world came the stuttering beginning of a feedback loop between the use of fire and the traversal of the subsurface—a self-augmenting circuit that has yet to abate. The ores that were fed into the fiery furnaces of the metalworker were extracted from its subterranean seams in a series of essentially fire-dependent operations. 'Fire-setting'—the cracking of rock through exposure to high heat followed by quenching—was the early miner's preeminent means of entering a world composed of solid, lithic layers. As Pyne elaborates: 'Prospectors burned over hillsides to expose rock. Miners relied on fire to tunnel, to smelt, to forge …They had to crush and process as much (ore) as possible on site, and nearly every stage demanded fire' (2001: 131). In turn, some of the ore pried from the depths returned as rock-hewing implements, tools harder than stone that helped fulfill rising demands for more metallic ores (Clark, 2021).

The key to this human adventure in verticality was the control and intensification of fire in robust, purpose-built chambers (Clark, 2021; 2022). Drawing both on the work of Gilbert Simondon and metallurgical histories, Deleuze and Guattari have helped us to grasp the significance of the metalworker's use of heat and a series of well-timed operations to coax material across a threshold into a novel organizational state. But there is a more general sense, taking insights from Henri Bergson, in which they speak of processes of involution, a kind of enfolding through which a little of the forces, properties, and potentialities bound up in any stratum can be extracted and introduced into the more flowing, non-stratified regions outside of the strata (1987: 238, 46–7).

We can conceive of the potter's kiln and the metallurgist's furnace, in this sense, as a literal involution of the geopower of the Earth: an enfolding of the material-energetic dynamism of the geological strata into the critical zone, or outer Earth System. In this context, it's worth recalling Deleuze and Guattari's warning about the inherent risks that attend the traversal of strata and any unbinding of their constitutive elements. This injunction seems informed in part by their reading of metallurgical histories, which often dwell on the dangers of trafficking between surface and the subsurface and on the practices through which traditional miners and metallurgists prepared themselves for this risky negotiation (Clark and Szerszynski 2023). In the words of R.J. Forbes, a key reference for Deleuze and Guattari, 'the smith grew to form a special social type encumbered with religious rites and rituals, endowed by popular feeling with magical potencies in many directions' (1950: 62). Historian Mircea Eliade adds that 'subterranean life and the spirits reigning there are ... a domain which by rights does not belong to man,' while noting the 'magico-religious power' associated with the transmutation the materials sourced from subsurface (1978: 56, 79).

To an even greater extent than open field burning, the use of fire to transmute earthy materials serves as an articulation between disparate strata or domains of existence, connecting past and future, the corporeal and the incorporeal, the sunlit outer world and the lightless subsurface. If it requires generations of trial and experimentation to acquire the necessary practical skills, so too has this traversal traditionally called for the cultivation of affective-embodied, ethical, and metaphysical comportment. But as I suggest in the following section, neither the fiery encounter with the soil nor with the stratified lithic layers of the Earth's crust exhausts the human capacity to intervene in planetary processes.

2.5 Fire and the Inner Earth

At the close of the twentieth century physicist Hans Joachim Schellnhuber described the Earth system as 'one single, complex, dissipative, dynamic entity, far from thermodynamic equilibrium' (1999: c20), a perspective that helped lay the foundations of the Anthropocene hypothesis. But as some geoscientists remind us, there is a second Earth system: the vast, unlivable world of superheated rocky matter beneath the planet's crust (Lenton 2016). Propelled by radioactive decay and residual heat from the accretion of the planet, the viscous rock of the inner Earth churns in massive slow-moving convection cycles, current models propose, punctuated by chaotic upwelling plumes of even hotter rocky material.

Anticipated by the tectonic plate hypothesis of the late 1960–early 1970s, contemporary solid Earth geophysics connects up the immense, slowly migrating slabs of rock comprising the Earth's crust and the overturning mantle beneath into a 'unified lithic-convective mantle system' (Coltice et al. 2017). Just as the Earth is our solar system's only fire planet, so too is it the only one of the rocky planets that has retained enough inner heat to drive an active plate tectonic-convective mantle system (Zalasiewicz 2008: 14–18). As many geoscientists insist, it is not simply the dynamics of the outermost Earth system that set the conditions for the habitability of our planet, but the evolving relationship between inner and outer Earth.

By way of developments in remote sensing, growing computational power, and novel data-sharing practices, scientists now have a much more detailed, historical understanding of this relationship than they did a decade and half ago. High-resolution '4D' modeling of feedbacks between tectonic plate motion and the changeable planetary interior allows researchers to track not only major events like the cyclical assembly and break-up of supercontinents but also finer-grained processes such as extrusions of lava, uplift, and rifting of landmasses, climate and ocean circulation change, sea level fluctuation, and the evolutionary pathways and dispersals of biological life (Zahirovic et al. 2019).

While no one is suggesting that anything lives beneath the crust, there is growing evidence that 'conveyor belts' of subducting plate material pull sedimented organic matter deep into the subcrustal Earth. In this way, over very long timescales, biological life has at least some impact on the composition of the planet's interior (Plank and Manning 2019), which must now include the possibility that human influence on outer Earth systems will eventually leave faint traces in the inner Earth. Moreover, human agents seeking to tap into geothermal energy sources have accidentally drilled into magma bodies deep in the Earth several times in recent years (Clark et al. 2018). Described by geoscientists as 'the first direct access to the magmatic environment of Earth' (ICDP 2017), we might tease out this observation to acknowledge that in the approximately four-billion-year span of life on Earth, this is the first time any living being has established sustained contact with the *in-situ* forces of the planetary interior.

That said, encountering upwelling magma stalled in crustal rock is not the same as accessing the Earth's subcrustal mantle layer. But we have already touched upon another means, indirect but more prevalent, through which skilled human actors have engaged with thermal processes constitutive of interior Earth. While I spoke of analogies between human high technics and metamorphic geology in the previous section, we can push this further—into the realms of igneous geology. So, when metallurgists concentrate and purify metallic ores by pushing them through solid–liquid-solid phase transitions, we might view them as effectively reproducing the transformations that occur when viscous rock from the mantle rises in the direction of the outer Earth—first melting then subsequently re-solidifying as it stalls in cavities in the crust or after it erupts (as lava) across the planetary surface (Clark 2022).

As metallurgist JE Rehder has noted, furnace temperatures of around 1200 °C first attained by artisans thousands of years ago are at the upper end of the heat of lava in active volcanic systems (2000: 54). While manipulating matter at this heat intensity is without precedent, in another sense, human use of controlled fire to replicate igneous processes has a considerable prehistory. Evidence from South African sites indicates that somewhere between 164–72,000 years ago, middle stone age artisans were deliberately heating stones beneath their hearth fires, most likely to improve their flaking and sharpening qualities but perhaps also because they were attracted to the blood-red coloration of some heat-treated rock (Brown et al. 2009). In the words of research team leader Kyle Brown: 'Here are the beginnings of fire and engineering, the origins of pyrotechnology, and the bridge to more recent ceramic and metal technology' (cited in ASU 2009).

We should recall that current evolutionary theory has hominins emerging in the vicinity of the volcanically active East African Rift—where the paleoanthropological record points toward stone-toolmakers coming to a very early appreciation of the workability of volcanic rock. As anthropologist Julien Favreau (2023: 1) notes 'hominins across Eastern Africa preferentially utilized igneous rock types followed by metamorphic and sedimentary lithologies mirroring the regional lithostratigraphy.' What the findings in South Africa imply is that at some point after they migrated away from the volcanic landscape of the Rift Valley, hominin craftspeople learned how to use heat in a controlled way to change the properties of available sedimentary rock so that it behaved much more like igneous rock.

While Anthropocene science focuses our attention on the rocky strata-Earth system interface, and critical zone theory zooms into the slender sunlit province animated by life, it's worth remembering that our genus spent its formative first few million years in and around a vast crack in the Earth's crust—our primordial home being a landscape torn apart by huge plumes of magma rising from deep within the planet's interior (Chang et al. 2020). Recent evolutionary theory has begun to recognize how significant this volatile and variable landscape was for an increasingly bipedal but relatively defenseless primate, both for the shelter it provided and for the nutrient-rich, well-watered ecological mosaic landscape it offered (King and Bailey 2006).

Most likely it was here, in terrain periodically set ablaze by both lightning and lava, that ancestral humans first captured and tended fire. Fire controlled in the hearth and later intensified in the furnace, I have suggested, became a means by which skilled human practitioners—earthly multitudes who convened around high heat—reproduced a tiny fraction of the igneous forces of the inner Earth. Emerging at a site of exceptionally active traffic between the mantle layer and the Earth's surface, our genus and species learned how to enfold some of the forcefulness of the planet's interior—and began to act as a peculiar kind of hinge or articulation between the thermal regimes of the inner and outer Earth.

Already in the last section we touched upon the awareness of high-heat-wielding craftspeople that mediating between distinct earthly regions was a dangerous wager. Historical accounts suggest fiery mishaps could be deadly for artisans and frequently left villages or city quarters in smoking ruin. In numerous 'Old World' cosmologies, metallurgists were incited and watched over by powerful deities, most often the same gods or spirits who presided over volcanoes and underground fire (Clark et al. 2018). But the later scaling up of metallurgical transmutation and the rise of new heat engines both intensified the human hinging together of inner and other Earth and rendered it commonplace. Deep within the internal combustion engines that still power most of planet's 1.5 billion motor vehicles, temperatures routinely exceed 2500°C, which is around the midway heat of the Earth's mantle layer. While we have lately learned to consider its impacts on the atmosphere, it is much rarer to think of this activity as a kind of traffic between inner and outer Earth—for all the recent efforts by geoscientists to weld these two planetary domains into a single dynamic system.

2.6 Future Fire and Planetary Care

Viewed through a geologic lens, fire applied to the open field leaves its mark in sedimentary processes, the baking of clay recapitulates the formation of metamorphic rock, while metallurgy and its precursors function as microcosms of igneous geology. Each of these ecologies of practice, I have suggested, utilizes high heat to intervene at a juncture between the layers or regions of the Earth. The burning of biomass is a means of manipulating the meeting place of above-ground life, atmosphere, and soil; introduced into rocky voids and enfolded into matter-transforming chambers, concentrated fire enables a traversal of the lithic strata that comprise the Earth's crust; further intensified and applied to rock and ores, concentrated flame serves as a proxy negotiation between the inner and outer Earth.

Just as the component parts of the Earth interact with each other, so too are these fiery practices overlapping and mutually informing. Because of the shared way they situate themselves in relation to specific structures and processes of the Earth, those who work with high heat—as with other earthly multitudes—are apt to appreciate each other's skills and knowledge, even across continents and through centuries. Moreover, we should be wary of assumptions that there is a clear succession, a simple step ladder from one use of fire to another. Stewart Brand's evocation of unfolding god-like powers, together with the Anthropocene scenario of novel planetary threats suggests a directional path, a telos of human world-altering powers. While this may hold true extensively—as an aggregation of brute force—when we take an evolutionary perspective, it is striking how early ancestral humans were at least nascently utilizing flame to traverse each of the major structural divisions of the Earth. It is as if, as a creature who emerged in and around the Great Rift of East Africa during bouts of volcanic activity and shifting climate, the niche of the genus *Homo* has always involved negotiating the Earth in '4D'—as contemporary modelers put it (Zahirovic et al. 2019: 73).

Playing with fire, it might be said, is what we do. By virtue of being an exothermic reaction, fire is always potentially runaway: with its capture comes a degree of risk unprecedented for biological beings. 'What it granted in power,' intones Pyne, 'anthropogenic fire demanded in responsibility' (1997: 322). Given that the hominin embrace of fire far exceeds the evolutionary span of *Homo sapiens*, we should take Isabelle Stenger's point about obligation as 'the *primordial* fact for an ecology of practice' quite literally (my italics). As Stengers's notion of practice entails, as Szerszynski and my earthly multitudes imply, and as much Indigenous fire use encapsulates, experimentation and care belong together. In this regard, it's worth remembering that hominins are creatures who are primordially obligated to a certain intensity and duration of caregiving by virtue of the exceptional dependency of our infants (Clark and Whittle 2023). Perhaps only a being with such extravagant experiences of responsibility would be capable of the amount of care required to sustain and nourish flame: to *tend* fire under wildly variant conditions, in vastly different environments, across multiple generations.

Steffensen's observation that '[f]ire is a commitment that goes on forever into the future' (2020: 87) is an entreaty to keep experimenting, to extend care for country even, or especially, as climate change dramatically raises the degree of danger. While lessons have been learned about cautious experimentation with open field burning, the matter of how best to use fire to negotiate other planetary gradients and boundaries cries out for attention. Alongside the now well-known consequences of exhuming and combusting fossil fuels, the environmental impacts of extracting many minerals from the subsurface are so devastating that some scientific-technical experts insist that mine tailings must be meticulously managed *in perpetuity* (Clark and Rickards 2022). In a broader sense, what we are seeing in western (and increasingly globalized) geoscience—associated in particular with climatology, Earth system science and the Anthropocene hypothesis—is the growing acknowledgment that injudicious human intervention in and around critical planetary junctures will have consequences that are effectively permanent. More than a matter of leaving behind sedimented and eventually stratified traces, this involves impinging on the planet's own self-organizing capacities, such that impacts rebound, amplify and cascade through tightly coupled Earth systems.

It seems that where western science-orchestrated thought and practice falls so fearfully short is precisely where a wider world of earthly or fiery multitudes is at their strongest—in the lore, rites, and observances that help safeguard operations at risky junctural zones. So routine and extensive is our trafficking across the great structural divides of our planet that it frequently withdraws from sight and awareness. And even the combustion of fossil hydrocarbons—which *has* been painstakingly rendered visible and subjected to contestation—eludes effective global regulation.

As fire creatures welded to a fire planet that is beginning to fume and rage to unfamiliar rhythms, we will be obliged to learn new heat-centered skills, practices, and strategies. So too, as we pass across borders and thresholds into unknown terrain—which will include confronting conditions that cannot be ameliorated—many of us will find ourselves having to improvise ethically, existentially, even cosmologically. But we shouldn't forget that our experiments are only a small part of the Earth's self-exploration, our own fiery probing of stratified matter-energy but a variation on the way this planet negotiates its own thermal gradients and boundaries.

Acknowledgements For opening up this conversation, I would like to thank Olivier Labussière and the rest of the organizing committee of the *Sols, Sous-sols dans la Transition Socio-écologique* conference, 9–11 June 2021. The author declares that there are no competing interests in relation to this chapter.

References

Arènes. A., Latour, B. & Gaillardet, J. (2018). Giving depth to the surface: An exercise in the Gaia-graphy of critical zones. *The Anthropocene Review*, *5*(2), 120–135. https://doi.org/10.1177/2053019618782257

Arizona State University (ASU). (2009). Early modern humans use fire to engineer tools. *ASU Now*. https://news.asu.edu/content/early-modern-humans-use-fire-engineer-tools

Australian Government (2011). Australian heritage database: Wilgie Mia aboriginal ochre mine. https://www.environment.gov.au/cgibin/ahdb/search.pl?mode=place_detail;place_id=106044

Bachelard, G. (1964). *The psychoanalysis of fire*. Routledge & Kegan Paul.

Black, M. P., & Mooney, S. D. (2007). The response of aboriginal burning practices to population levels and El Niño-Southern oscillation events during the mid- to late-Holocene. *Australian Geographer, 38*(1), 37–52.

Bowman, D. M. J. S., O'Brien, J. A., & Goldammer, J. G. (2013). Pyrogeography and the global quest for sustainable fire management. *Annual Review of Environment and Resources, 38*, 57–80.

Brockman, J. (2009). We are as gods and have to get good at it. http://www.edge.org/conversation/stewart_brand-we-are-as-gods-and-have-to-get-good-at-it

Brooke, J. (2014). *Climate change and the course of global history: A rough journey*. Cambridge University Press.

Brown, K., Marean, C., Herries, A., et al. (2009). Fire as an engineering tool of early modern humans. *Science, 325*(5942), 859–862.

Chang, S. J., Kendall, E., Davaille, A., & Ferreira, A. M. G. (2020). The evolution of mantle plumes in East Africa. *Journal of Geophysical Research: Solid Earth, 125*, e2020JB019929. https://doi.org/10.1029/2020JB019929

Clark, N. (2018). Fire: Pyropolitics for a planet of fire. In K. Peters, P. Steinberg, & E. Stratford (Eds.), *Territory beyond terra* (pp. 69–85). Rowman & Littlefield.

Clark, N. (2021). Vertical fire: For a Pyropolitics of the subsurface. *Geoforum, 127*, 364–372.

Clark, N. (2022). Planetary cities: Fluid rock foundations of civilization. *Theory, Culture & Society, 39*(2), 177–196. https://doi.org/10.1177/02632764211030986

Clark, N. (2023). Exchanging fire: A planetary history of the explosion. In A. Bashford, E. Kern, & A. Bobbette (Eds.), *New earth histories: Geo-cosmologies and the making of the modern world* (pp. 209–224). University of Chicago Press.

Clark, N., Gormally, A., & Tuffen, H. (2018). Speculative volcanology: Violence, threat and chance in encounters with magma. *Environmental Humanities, 10*(1), 273–294.

Clark, N., & Rickards, L. (2022). An Anthropocene species of trouble? Negative synergies between earth system change and geological destratification. *The Anthropocene Review, 9*(3), 425–442. https://doi.org/10.1177/2053019622110739

Clark, N., & Szerszynski, B. (2021). *Planetary social thought: The Anthropocene challenge to the social sciences*. Polity.

Clark, N. & Szerszynski, B. (2023). Planetary technics, earthly spirits. In S. Bergmann, K. Rigby, & P. M. Scott (eds) *Religion, Materialism and Ecology*. Routledge.

Clark, N. & Whittle, R. (2023). For a paleogeography of childcare: Infant-carrying technics on a dynamic planet. *EPF: Philosophy, Theory, Models, Methods and Practice*. https://doi.org/10.1177/26349825231200606

Coltice, N., Gérault, M., & Ulvrová, M. (2017). A mantle convection perspective on global tectonics. *Earth-Science Reviews, 165*, 120–150.

Deleuze, G., & Guattari, F. (1987). *A thousand plateaus: Capitalism and schizophrenia*. University of Minnesota Press.

Dick, S. (2019). *Classifying the cosmos: How we can make sense of the celestial landscape*. Springer.

Eliade, M. (1978). *The forge and the crucible*. University of Chicago Press.

Favreau, J. (2023). Sourcing Oldowan and Acheulean stone tools in Eastern Africa: Aims, methods, challenges, and state of knowledge. *Quaternary Science Advances, 9*, Article 100068. https://doi.org/10.1016/j.qsa.2022.100068

Forbes, R.J. (1950). *Metallurgy in antiquity*. E.J. Brill.

Granjou, C., & Meulemans, G. (2022). Bringing soils to life in the human and social sciences. *Soil Security., 10*, Article 100082. https://doi.org/10.1016/j.soisec.2022.100082

Hunt, W. (2019). *Underground: A human history of the worlds beneath our feet*. E-book. Simon & Schuster.

International Continental Scientific Drilling Program (ICDP) (2017). *Krafla Magma drilling project workshop*. https://www.bgs.ac.uk/icdp/blogs.

Isabella, J. (2013). The Caveman's home was not a cave. *Nautilis*. https://nautil.us/the-cavemans-home-was-not-a-cave-234672/

King, G., & Bailey, G. (2006). Tectonics and human evolution. *Antiquity, 80*, 265–286.

Kimmerer, R. W., & Lake, F. K. (2001). The role of Indigenous burning in land management. *Journal of Forestry, 99*(11), 36–41.

Labussière, O. (2021). A coalbed methane 'stratum' for a low-carbon future: A critical inquiry from the Lorraine Basin (France). *Political Geography, 85*, Article 102328. https://doi.org/10.1016/j.polgeo.2020.102328

Lake, F. K. & Nelson, A. (2022). Expert Q&A: Frank K. Lake. *Biohabitats,xx*(2). https://www.biohabitats.com/newsletter/fire-on-the-landscape/expert-qa-frank-k-lake/

Latour, B. (2014). Some advantages of the notion of 'Critical Zone' for geopolitics. *Procedia: Earth and Planetary Science, 10*, 3–6.

Leach, M., Fairhead, J., & Fraser, J. (2012). Green grabs and biochar: Revaluing African soils and farming in the new carbon economy. *Journal of Peasant Studies, 39*(2), 285–307.

Lenton, T. (2016). *Earth system science: A very short introduction*. Oxford University Press.

Lövbrand, E., Beck, S., Chilvers, J., Forsyth, T., Hedrén, J., Hulme, M., Lidskog, R. & Vasileiadou, E. (2015). Who speaks for the future of Earth? How critical social science can extend the conversation on the Anthropocene. *Global Environmental Change, 32*(Supp. C), 211–218.

Lyons, K. M. (2020). *Vital decomposition: Soil practitioners and life politics*. Duke University Press.

Plank, T., & Manning, C. (2019). Subducting carbon. *Nature, 574*, 343–352.

Puig de la Bellacasa, M. (2017). *Matters of care: Speculative ethics in more than human worlds*. University of Minnesota Press.

Pressler, Y., Moore, J. C., Cotrufo, M. F. (2019). Belowground community responses to fire: meta-analysis reveals contrasting responses of soil microorganisms and mesofauna. *Oikos, 128*, 309–327. https://doi.org/10.1111/oik.05738

Pyne, S. J. (1994). Maintaining focus: An introduction to anthropogenic fire. *Chemosphere, 29*(5), 889–911.

Pyne, S. J. (1997). *World fire: The culture of fire on earth*. University of Washington Press.

Pyne, S. J. (2001). *Fire: A brief history*. University of Washington Press.

Rehder, J. E. (2000). *The mastery and uses of fire in antiquity*. McGill-Queens University Press.

Santín, C., & Doerr, S. H. (2016). Fire effects on soils: The human dimension. *Philosophical Transactions of the Royal Society B, 371*, 20150171. https://doi.org/10.1098/rstb.2015.017

Schellnhuber, H. J. (1999). "Earth system" analysis and the second copernican revolution. *Nature, 402*(6761), C19–C23.

Smil, V. (2006). *Energy: A beginner's guide*. Oneworld.

Steffensen, V. (2020). *Fire country: How Indigenous fire management could help save Australia*. Hardie Grant Travel.

Stengers, I. (2005). Introductory notes on an ecology of practices. *Cultural Studies Review, 11*(1), 183–196.

USDA. (2019). Faces of the forest service: Meet Frank Kanawha lake https://www.fs.usda.gov/faces-of-the-forest-service/meet-frank-kanawha-lake

Wellings, S. (2016). Man-made metamorphic rocks. http://all-geo.org/metageologist/2016/03/man-made-metamorphic-rocks/

Zahirovic, S., Salles, T., Müller, R. D. et al. (2019). From Paleogeographic maps to evolving deep-time digital earth models. *Acta Geologica Sinica, 93*(supp.1), 73–75.

Zalasiewicz, J. (2008). *The earth after us: What legacy will humans leave in the rocks?* Oxford University Press.

Zorich, Z., (2014). Early humans made animated art. *Nautilus*. March 27. https://nautil.us/issue/11/light/early-humans-made-animated-art

CHAPTER 3

Photographing Glaciers as They Are

Olivier de Sépibus Thomaïdis and Olivier Labussière

Olivier Labussière: What is Your Relationship with the Mountains?

Olivier de Sépibus Thomaïdis: My parents always joked that I knew how to ski before I could walk! I grew up in Gap, in the Écrins massif. In the 1970s, the development of ski resorts coincided with a minor global cooling, which serendipitously benefited them. At the time, I lived in Tallard in the Hautes-Alpes, 600 meters above sea level, where winters brought about 40 centimeters of snow. My father nurtured a deep appreciation for snow culture in me. I started by joining him on less challenging climbs, and to elevate my skills, I joined the French Alpine Club. I was immersed in mountaineering stories from a young age. The mountains

O. de Sépibus Thomaïdis (✉)
Die, France
e-mail: olive@olivier-de-sepibus.com

O. Labussière
Pacte Social Sciences Research Center, University Grenoble Alpes, CNRS, Sciences Po Grenoble, Grenoble, France
e-mail: olivier.labussiere@cnrs.fr

© The Author(s), under exclusive license to Springer Nature Switzerland AG 2025
O. Labussière et al. (eds.), *Back to the Ground*,
https://doi.org/10.1007/978-3-031-88888-5_3

are inseparable from the stories we tell about them. Mountaineers often turn to writing and storytelling as they are frequently the sole witnesses of their adventures. Walter Bonatti, for example, was not only an exceptional climber but also a gifted writer who vividly brought his climbs to life. Lionel Terray's *Conquistadors of the Useless* (originally published in 1961 under the title *Les Conquérants de l'inutile*) was a seminal work. By contrast, Louis Lachenal was a great mountaineer, but he struggled to effectively convey his experiences. These stories captivated me, fueling my aspiration to become a high mountain guide. Then, around 1987, I nearly lost my life during an ascent of a serac—a huge block of glacial ice on a glacier. That fall prompted a period of deep reflection and fear, leading me to stop mountaineering altogether.

Photography Entered Your Life Quite Early, Didn't it?

Yes, thanks to my father. He ran a photographic film development business in Gap (Southern France) and served the entire Rhône Valley and the French Riviera. So, I grew up surrounded by photographic developer! We had rolls of film and cameras all over the house. I was free to take as many photos as I wanted. When I was 13, my father taught me the basics of photography, and I had access to a small black-and-white lab. Storytelling and photography are two crucial mediums, as the great alpinist Gaston Rébuffat knew well. He created beautiful books with stunning photos, including *The 100 Finest Routes in the Écrins* (1975) and *The 100 Finest Routes of the Vanoise* (1977). He worked with skilled photographers who captured him in action on the mountains.

And you Chose to Become a Professional Photographer

As a teenager, I dreamed of doing photojournalism for magazines and traveling. As soon as I came of age, I began to travel. At that time, there wasn't formal schooling for the type of photography I was interested in. There were programs for advertising photography, but nothing for photojournalism. I learned in the school of the streets. I quickly began working for a climbing magazine, *Grimper Magazine*, and loved it. I was right in the action, documenting 8A and 8B climbs and enjoying my time among the climbers. Later, I tried social documentary photography for Parisian magazines, but it didn't really take off. I published a book on the "Chibanis" of the Belsunce district in Marseille with the poet Habib Tengour, titled *Retraite* (Tengour & de Sépibus, 2004). Around the same time, I discovered Land Art. I was captivated by artists like Richard Long and Andy Goldsworthy, who work in and with nature. It was then that I realized I had something to contribute to this field. In 2004, I revisited the Glacier Blanc—located in the Massif des Écrins (French Alps), and the shock was profound! It had receded far beyond where I'd last seen it in 1986.

What Happened that Day at the Foot of the Glacier Blanc?

That day, I was alone. It felt a bit like a pilgrimage. For those of us from the Hautes-Alpes region, the Glacier Blanc is THE glacier—it is cloaked in an almost mystical aura, with the towering Écrins ridge above. I was also returning to a place of my youth, a place that had shaped me. Alpinism is a serious pursuit. It might sound like a cliché, but the rope team must hold together; our lives are interwoven, which is essential. When I reached the spot where the glacier's front had been 15 years earlier, it was gone. What happened inside me, what "collapsed" inside me… was that the landscape was disfigured. Its main feature, the glacier, the star around which the legend of the Alps was built, is vanishing. It was an aesthetic shock!

This Aesthetic Shock Highlights Your Intimate Relationship with Glaciers.

I was destabilized. The mountains I see are no longer the ones I hold within me, the ones I've internalized. I realized that I am permeated by them. There are many ways to describe a relationship with nature—"being in contact," "being connected," "being immersed"—but for me, the notion of permeation seems most fitting. A submerged body can remain unaffected, yet the idea of permeation suggests that I am as infused by the environment as I infuse it. We are shaped by the environment of our childhood, and there's no changing that. What took me a long time to realize through photography is that my work isn't just about the act of looking. It's a very tactile relationship. In this respect, my relationship with nature isn't the one described by the anthropologist Philippe Descola (2013), who argues that Western society maintains a relationship of contemplation and distance with the landscape. I find more resonance in the sensitive geography of Augustin Berque (1986, 2009, 2019).

This Turmoil Also Calls into Question Your Relationship with Imagery. You Revisit the Early Representations of the Alps.

Confronted with the glacier's retreat, I realized I could no longer photograph the mountains in the conventional, illustrative way I had before. I felt compelled to return to the basics, to do my research. I discovered the works of Claude Reichler, a Professor of Literature at the University of Lausanne: *Les Alpes et leur imagier* (2013). He has extensively studied how the Alpine landscape emerged in cultural representations during the eighteenth and nineteenth centuries. Through these texts, I encountered the first images produced by artists like Caspar Wolf in 1788–89. This was around the time of Rousseau's greatest influence, when the English bourgeoisie first became attracted to the Alps, treating it as a kind of *Terra incognita*. Both scientists and artists were drawn to these new territories, creating images that no longer depicted the mountains as menacing and haunted by dragons but rather as aesthetically sublime.

What do these Images Say About our Cultural Relationship with the Mountains?

Through these images, I've noticed a shift in scale and perspective. In pre-seventeenth-century representations, the scale ratios are peculiar: Humans are depicted as very large, and the mountains as very small. The perspective is commanding, demiurgic—it's all about man, the supreme creation. With the shift toward scientific descriptions, the proportions flipped: humans became minuscule against the vastness of the mountains. Western man stages himself in a hostile environment, yet his smallness is also his greatness. Visual codes were established during this time, and we continue to live with them today. Now, the Anthropocene has prompted a re-evaluation of this scale and perspective. We are confronted with the contradictory reality that humans are both "greater than" and "smaller than" the terrestrial; humans are a geological force yet also vulnerable beings. The traditional codes of representation are shattered. The question of humans' place in nature is becoming a central issue.

And yet, we Continue to Rely on an Outdated Collection of Mountain Imagery

The postcards of high mountains still sold in tourist shops feature images from the 1980s. For example, the glaciers in the Écrins have significantly diminished since then. The images in circulation are outdated and no longer reflect current realities. It's challenging today to keep pace with this transformation and to create an aesthetic that matches this new reality. The sense of catastrophe is intensified because it's impossible to envision a desirable future without adequate representations. When I visit the mountains, it's heart-wrenching to see how alpinists, refuge keepers, glaciologists and hikers—both professionals and amateurs—are mourning this loss. It's devastating. We lack alternative representations.

We Need to Imagine New Visual Codes to see the Mountains as they are

As I delve into the history of Alpine representation, I identify and focus on three visual codes that define what mountains mean to me: first, the interplay of shadow and light that, through intense, almost violent constraints, evokes a sense of the sublime; second, a perspective that lends the mountain depth and vastness; and third, the relationships of scale. This last aspect is particularly important to me. Typically, mountain depictions include tiny human figures to add drama to the scene. The smaller the human figure, the braver they appear, daring to lose themselves in this wild and hostile vastness. I intend to turn these visual codes inside out. For example, I've realized that I don't want the relationship of scale to be playful. With the Anthropocene, the scale relationship has become destabilized, disordered and troubled. It felt right to photograph the mountains without human presence, allowing my images to also question the space humans occupy in these landscapes.

Evolving Your Photographic Practice also Means Forging a New, Sensitive Connection with the Mountains.

Back in 2004, I was unsure about what I might find, what I was looking for or the kind of photos I would end up taking. I started capturing images without any specific aim. Every year, I head to the mountains in late August when the glaciers are most accessible—a time that's typically unconventional for photography. Over a decade, from 2007 to

2017, my work explored a dialogue with chaos. Initially, I was drawn to photographing rockfalls—elements I saw as the overlooked features of mountain landscapes. My fascination lay in framing these chaotic scenes. Capturing a rockfall is somewhat like bringing order to what is typically perceived as disorder. Today, as we navigate a period of upheaval, there's a pressing need to create new narratives, as Baptiste Morizot and Nastassja Martin (2018) suggest. We're stepping back into a mythical era where artists play a crucial role.

In 2022, you Set Out on a Tour of the Alps to Visit the Glaciers. What did you See?

Amid the backdrop of COP 21 taking place in Paris, interest in my photos grew. I received a grant from the French National Centre for Visual Arts in 2021 and support from the Auvergne-Rhône Alpes regional culture department in 2022 to embark on a month-long Alpine tour starting in Tyrol. This support allowed me to expand my collection of images. The summer of 2022 turned out to be one of the hottest on record. The glaciers accumulated more rockfalls, dust, Sahara sands and pollutants on their surfaces than snow. I began my journey in Savoie, near the Évettes Glacier. Then, I crossed into Italy, below Monte Rosa, near Macugnaga—an absolutely incredible place where rockfalls are a constant presence. Next, I visited Austrian Tyrol, where the mountains are less steep. The glaciers seemed less dynamic and looked more like melting ice cubes. Next up was Switzerland, where I headed to the Bernina range. There, I found impressive "meringue-like" glaciers that still had a dynamic appearance but were retreating far more rapidly than models had predicted. On the Italian side, I discovered a glacier spilling into a lake. Then, I ventured to an extraordinary place, the Lauteraargletscher Valley in Switzerland. This valley quickly becomes wild and austere and is home to some of the largest Alpine glaciers, including the nearby Aletsch Glacier. The final leg of my tour took me to the Aiguille du Midi, where I camped at the Plan de l'Aiguille. In the morning, as I was nearing the end of my film roll, there was a massive collapse under the Aiguille du Midi. For 20 minutes, we were engulfed in a cloud of orange dust. I chose not to take any pictures, reflecting instead on how my work continues to evolve with the landscape—capturing both its emergence and its collapse, as well as the stones now blanketing the ice.

You Returned From Your Tour of the Alps with Photographs that Depict a Mountain Undergoing De-Stratification.

When I arrived in Chamonix, the last stop on my tour, it had just snowed. The summit of Mont Blanc was pristine white, contrasting sharply with the lower parts that retained the dust and pollutants from the summer. The snow-rain line was strikingly visible. It was an incredible sight. A few years ago, a magazine rejected my photos, claiming they didn't want a "dirty mountain." In Alpine iconography, the association with whiteness is highly valued. We still cling to visual codes that no longer reflect the current state of the high mountains. It takes time to mourn the loss of the white mountains and the glaciers… When I photograph moraine remnants where the glacier front once stood, it isn't obvious to most people. They only see stones. It's challenging to photograph things as they are, to truly capture the remnants of a glacier for what they are. And you see, I've even created images that are, to me, violent. There's a brutality in what I capture, which I then struggle to exhibit. When I see the remnants of a glacier, dripping and hanging between rocky spurs… that image, to me, is like the crucifixion. It's Christ on the cross. We're in a time of lamentation and mourning in the mountains near the glaciers.

Glaciers also Show us that When One Thing Disappears, Something Else Emerges. You Call this an "Undesired Landscape."

The landscape changes radically, and both its identity and its representations must also be updated. This presents a collective challenge because the emerging landscape—this landscape of stone expanses—isn't what was expected. It's not desired. I feel a responsibility to undertake this photographic work because, to grasp new situations, we need representations, images, narratives and new imaginations. Even if undesired, this is the landscape our children will come to know, and it must be represented. This requires us to open up to new emotions. In a landscape of stones, we still struggle to see what is emerging. For instance, in Austrian Tyrol, the retreat of the Ferner glacier has revealed a rock smoothed by ice over millions of years. This rock is now visible for the first time. For me, this relationship with unveiling is profound; it moves me. We need to learn to see it for what it is.

The De-stratification of the Mountain also Influences our Language. We Struggle to Find the Right Words for what we See

Today, we still struggle to accept what the high mountains are becoming. There's little human adventure in these newly exposed spaces, and the task of naming such places is ongoing. For example, how do we name a place that has lost its glacier, like the base of the Clôt des Cavales glacier? This reshaping of language is something we're grappling with. Inspired by my images, Claude Reichler (2024) coined the concept of "effigy." The term describes what remains after something has disappeared. Naming what the glacier reveals is challenging but essential for sharing our experiences. We need new names and words for things that previously went unnamed. My photographs inspired Reichler to come up with the term "effigy" as I'm one of the first to show this new face of the *undesired* mountain.

How do you Present These Photos to the Public?

I envision my exhibition as a wander at 3,000 meters altitude. I make it clear where each photo was taken by using placards placed at a distance, ensuring the location is not the focus. Each photo is uniformly formatted to 1 meter by 1 meter. By maintaining this uniform size, the rhythm isn't dictated by the display setup but plays out within the images themselves—through the shimmering colors of the rocks and glaciers, as well as the rhythms of form. *Earth and Reveries of Will*, the formation of an inner image, that's Bachelard! It's a musical relationship with the world. From a distance, the images appear almost abstract, but their high definition invites viewers to "enter" into the textures and colors. I photograph *from glacier height*, to show them as they are.

In front of my photos, I encourage viewers not just to look but to scrutinize, to detect clues and traces as a climber studies a cliff before an ascent and thinks, "I'll go up here, then here." This way of seeing focuses on seemingly mundane spots. To scrutinize is to find the line, to search for the path—it goes here, there, turns right… This activity of scrutiny sharpens our vision and draws us into a dynamic mountain, now characterized by flows and collapses.

Sharing These Images is a Challenge, it Calls for Exploring new Modes of Reception

I'm still struggling to find the right venues and media for displaying these images, although things are gradually changing. I do sometimes publish these images, but the tourism industry continues to crave the whiteness of the mountains. The entire Alpine arc is entrenched in this cultural attachment to glaciers and the prevailing visual codes. Meanwhile, I navigate a mountain undergoing de-stratification, focusing on emerging images and narratives. It's not easy. As an artist seeking expression, I've proposed to a festival to display about 10 photos in the street, without commentary. This approach is inspired by Ernest Pignon-Ernest, who exhibits his Pietàs and figures of suffering in public spaces. I aim to set up a display that reflects the Earth's suffering out on the streets; it seems like a fitting way to convey what's happening.

The Ongoing Disappearance of Glaciers is Continually Reshaping Your Artistic Approach, Leading you to Some Intense Drawing Over the Past Year

As the Alpine landscape transforms, with rock chaos and moraines growing more prominent, I've felt compelled to explore these stony expanses with drawings. I begin with a photograph, project it and then translate it onto paper using charcoal. This method brings me closer to the raw physicality of the landscape. To truly draw the world's surface, one must understand its core—it requires immersion. Without a deep, internal connection, it's impossible to reinterpret our relationship with the mountain as it now exists.

You're Exploring a Medium that Fosters a Deep, Inward Connection with the Glacier as a Subject

I photograph the glaciers as they are but also because they exist—they're landscapes that demand to be seen as subjects. I create portraits of them. Each image is both a call to the mountain and a response from it, opening a dialogue, a space where its personality unfolds and communication begins. My approach to treating glaciers as subjects stems from the deep, poignant connections I feel with them, where their disappearance feels almost as personal as a death. It's not a person, of course, but it's a profound loss that affects me deeply. I hope that these images can act as a bridge between humans and these natural entities, attributing to them an inner life. What's fascinating is the potential for legal recognition of these entities as subjects, similar to how certain cultures recognize rivers or forests as legal persons because of pre-existing profound connections. In Western society, it's a challenge for art and perception to carve out this path. As an artist, I delve into this inner realm and return to share the story of my encounters with the glaciers. From there, the law might begin to recognize this newly perceptible reality.

The reader may also be interested in visiting the Olivier de Sépibus website: https://olivier-de-sepibus.com/lang-2/?p=4d6f6e7461676e652064c3a96661697465/50617263652071752696c20636f756c6520656e206e6f7573206573720676c616369657273

Acknowledgements The authors declare that there are no competing interests in relation to this chapter.

References

Bachelard, G. (2002). *Earth and reveries of will*. Dallas Institute of Humanities and Culture. (Original work published 1948)

Berque, A. (1986). *Japan: Nature, artifice, and Japanese culture*. Pilkington Press Ltd.

Berque, A. (2009). The choretic work of history. *Semiotica, 175*, 163–176.

Berque, A. (2019). An enquiry into the ontological and logical foundations of sustainability: Toward a conceptual integration of the interface 'Nature/Humanity'. *Global Sustainability*, 2. https://www.cambridge.org/core/journals/global-sustainability/article/an-enquiry-into-the-ontological-and-logical-foundations-of-sustainability-toward-a-conceptual-integration-of-the-interface-naturehumanity/0EDFB1653EDF5A45A07C3D75EFAFEF2E

Descola, P. (2013). *Beyond nature and culture*. University of Chicago Press. (Original work published 2005)

Morizot, B., & Martin, N. (2018). Sur un destin commun des animistes et naturalistes face au changement climatique à l'Anthopocène. *ISSUE, Journal of Art and Design*. https://issue-journal.ch/focus-posts/baptiste-morizot-et-nastassja-martin-retour-du-temps-du-mythe-2

Reichler, C. (2013). *Les Alpes et leur imagier: Voyage et histoire du regard*. Savoir suisse.
Reichler, C. (2024). *Les glaciers des Alpes et la photographie: Dans la lumière de leur disparition*. Presses Universitaires de Rennes.
Tengour, H., & de Sépibus, O. (2004). *Retraite*. Le Bec en l'air.
Terray, L. (1961). *Les Conquérants de l'inutile*. Gallimard.

CHAPTER 4

The Underground in Science Fiction: A Revealing Playground? Using Science Fiction to Support Traditional Foresight

Gaëlle Deletraz, Mathilde Joncheray, and Delphine Montagne

List of Abbreviation

IPCC	Intergovernmental Panel on Climate Change
ESA	European Space Agency
NASA	National Aeronautics and Space Administration

G. Deletraz (✉) · D. Montagne
University of Pau and Pays de l'Adour, CNRS, TREE, Pau, France
e-mail: gaelle.deletraz@univ-pau.fr

D. Montagne
e-mail: delphine.montagne@univ-pau.fr

M. Joncheray
University Jean Jaurès, CNRS, LISST, Toulouse, France
e-mail: mathilde.joncheray@univ-tlse2.fr

© The Author(s), under exclusive license to Springer Nature Switzerland AG 2025

O. Labussière et al. (eds.), *Back to the Ground*,
https://doi.org/10.1007/978-3-031-88888-5_4

4.1 Introduction

Subterranean spaces, which are often overlooked, rendered invisible, or diminished in public debates, policies, and ecological concerns, are crucial to the functioning of our societies, economies, and ecosystem balances (Montgomery, 2015; Fortey, 1994). Facing future challenges and the necessities of ecological transitions essential for preserving the planet's habitability, it is beneficial to reflect on our relationship with the future while emphasizing anticipation and foresight. The role of imagination in this context is variable and evolving. Meant to inform political decisions and public action toward a desirable future, foresight has often proved (albeit in retrospect) to be of limited effectiveness, as its methods are frequently viewed as too "linear" (Adam, 2023). Considering that science fiction offers a unique framework for exploring the role of subterranean spaces, we propose an analysis aimed at identifying and understanding their imagined potentialities, interactions with the surface, and impact on future societies. This work, primarily conducted on a corpus of literary science fiction works, is exploratory. Though not exhaustive, this initial analysis reinforces the idea that this material is a valuable tool for envisaging alternative futures and enriching debates on transitions.

4.2 Prospective, Relationship to the Future, and Interest in the Imaginary

In its early days, foresight was envisioned as a fusion of will and imagination, merging rigor and creativity. Pioneers like Saint-Paul and Teniere-Buchot (1974 in Smida, 2004) conceptualized foresight as the art of "thinking the unthinkable," "inventing the future," and "wanting the possible," emphasizing the importance of overcoming taboos and daring to be original while adhering to certain norms to shape the future. This vision, inspired by figures like H.G. Wells and Gaston Berger, reflects the ambition to combine scientific exploration with narrative imagination (Polère, 2012). However, despite these intentions, foresight quickly reverted to more traditional methodologies, especially in France, where the pressure for scientific recognition favored quantitative and rational approaches over imaginative exploration (Henshel, 1981; Gimbert, 2004).

Centered on the use of historical data, economic or social models, and statistical simulations, the discipline has long favored an approach based on rationality and operability, mainly targeting macroeconomic issues (Smida, 2004; Scouarnec, 2008). Reinforced by a predominant scientific culture among foresight practitioners, there was an early recognition of a certain inability to ultimately challenge existing power structures, a tendency for foresight practitioners to conform to the interests and policies of the established order (Domenach, 1966).

Fifty years later, Groves (2017) essentially echoes the same observation, based on a more systemic analysis. He provides a detailed analysis of the combination of expertise and sociotechnical apparatuses that reinforce divisions of labor and existing process flows within the governance of planning, illustrating what has been called "planning cascades" (Owens, 2004 in Groves, 2017). According to him, such cascades result in allocating available time and energy to respond to fragmented and specific planning needs rather than reflecting more deeply on the nature of the strategic need. These practices turn the future into an abstract construct, frozen in specific scenarios that rely on quantifiable and often disembodied approaches to the future. This approach treats the future as a series of calculable and optimizable possibilities commonly used in infrastructure planning. They justify decisions based on demand forecasts or estimates of future risks. Drawing on Foucault, Groves links these approaches to forms of power, which contribute to creating social coalitions over different representations of the future, shape environmental conflicts by distributing power unevenly between different social groups, and forge the imagined futures that will become the "public things" on which public policies will later focus, including infrastructure governance (Groves, 2017).

Beyond the methods of foresight or styles of anticipation, it is more fundamentally the visions of the future and the multiplicity of societies' relationships to the future that impact and direct our capacity for anticipation and action. Various authors have sought to theorize this issue. Beck (2001) and Morin (2020), for example, through the theory of the "risk society" for the former and the "theory of complexity" for the latter, both emphasize the confrontation of contemporary societies with risks, uncertainties, complex and uncertain situations that affect the capacity of societies and individuals to plan for the future. Part of this deep malaise is now labeled as eco-anxiety (Clayton et al., 2017).

For Hartog (2003), we have been living under a "presentist" regime since the 1990s. This regime of historicity is marked by the prevalence of the present viewpoint, which overshadows both past and future perspectives. According to him, although we continue to look forward and backward, our focus does not move beyond a present that is multidirectional (Hartog, 2003). The present has become a horizon devoid of a future and a past, which generates on a daily basis the past and future it needs at the moment *t*. This regime values the immediate, the present, at the expense of the future (Lessault, 2004). Adam (2010) delves into this issue and demonstrates how the future is often treated by foresight as a free resource to be used in the present, on which much of the wealth and global domination of Western and Westernized societies is based.

After decades of warnings, the materialization of environmental changes has drastically shifted perceptions of these issues. Several environmental or resource-related planetary boundaries have been crossed (Steffen et al., 2015; Boutaud and Gondran, 2020), including those concerning the consumption of rare earths and precious metals (Vidalenc et al., 2022). These issues are considered on a global scale, in extended temporal frameworks, with a horizon extending to 2050 and beyond. They pose challenges of justice, such as equitably distributing the burdens of climate change, its impacts, and the costs of necessary investments.

This present reflects the ambitions and actions of our predecessors, whose visions have shaped our reality, now facing environmental and social challenges (Adam, 2010). Current generations thus inherit the costs of past decisions, caught between lamenting the damage caused and the necessity to rectify past mistakes (Jonas, 1995). For now, we generally take for granted that the future is created by us, in and for the present. The future is then seen as subject to our will, even though it does not always unfold as planned (Adam, 2010). Indeed, we maintain hope for a technology *yet to be developed*, capable of overcoming our limitations and correcting ongoing imbalances. These ideas of solutions (e.g., geoengineering, green technology, connected objects, artificial intelligence, post-humanism, etc.)—infinite repairs—are particularly prevalent among the elites (Meheust, 2012). They carry the risk that each technological correction introduces an increasing and unpredictable number of novelties (and, thus, disruptions), which are then passed on to future generations. This singular approach is ultimately a dead end.

According to Morin (2020), "*We must no longer continue on the road of 'development.' We need to change lanes, we need anew beginning.*" Adam (2010) also notes that it is no longer possible to view the future as a blank slate to be freely shaped, because the consequences of the futures projected by our predecessors, the "futures of the past," now impose themselves on the present, limiting our choices and options. The ownership of the futures we create comes with obligations and responsibilities (Adam, 2010). As owners and creators of the future, we become responsible for the outcomes of our future-creating actions. This responsibility also pertains to the ethics of the future developed by Jonas (1997). Thus, these reflections raise the question of the necessity to adopt a new approach more in line with contemporary realities.

To reform our visions of the future, Adam (2010) suggests distinguishing the future according to whether it represents a probable, possible, or preferable outcome. She also emphasizes the distinction between futures as sources of market values (more or less free in the present) and futures as a matter of collective care. By contrast, Groves (2017) invokes the notion of "style of anticipation." Unlike the data and model-based approach mentioned earlier, Groves proposes a style of anticipation based *on lived and affectively engaged futures*, focusing on experiences, emotions, and attachments related to specific places and communities. This approach encompasses futures as imagined and felt by individuals and communities in their daily lives, highlighting local concerns, stories, and values in specific places and communities. Adam (2023) directly calls for the use of imagination, experience, and existing knowledge (memory) to navigate uncertain futures.

Between sobriety in the use of resources and technoscience or geoengineering (carbon capture and storage), the underground plays a crucial role, even if its place in public debate is often obscured or confined to arenas of specialists and/or residents (Missaghieh-Poncet, 2023). The necessity of a transition compels us to reconsider—among other things—the dynamics, formations, and stability of the underground (Hawkins, 2020). Integrating the concept of "geological eras," via the Anthropocene, into the concerns of the political and industrial world is delicate. The timescales of soils and subsoils are incredibly long: Biogeochemical cycles like that of carbon, or the even less tangible cycles of rock and soil creation and their interactions with the atmosphere and oceans, are difficult to comprehend. The clash between these geological times and the "time of humans," and even more so with "political time," presents a significant challenge. This might require recourse to visions that extend beyond rationality and scientific aspects, drawing on shared imaginaries and cultural understandings closer to lived experiences to prepare for what may seem—at a given moment—"unthinkable."

From this perspective, Rumpala (2010), Girard and Gendron (2010) and Gendron et al., (2012) are among the first to signal the potential of science fiction for academic research. It acts as a revealer of our relationship to technology, others, and our environment, and it invites exploration of trajectories that are markedly different from observable trends. This interest is also evident in the field of public policy. The Strategic Analysis Center (Barreau et al., 2012), the European Space Agency (ESA), and, more recently, the French military with the RED Teams (Dussert, 2020) experiment with integrating science fiction material into their reflections. By enabling thought experiments (Rumpala, 2016), we uncover "*alternative worlds that the foresight practitioner might not have the audacity to imagine*" (De Jouvenel, 2016). Rumpala (2018) presented stimulating reflection paths on the question of Earth's habitability based on the analysis of a large corpus of science fiction.

Thus, in a context where there is a need to renew approaches to anticipation to better align with contemporary issues, communities involved in future scenario planning highlight the value of science fiction. Its ability to deconstruct our representations and to help correct the "myopia" that affects us is particularly valued. It is with this mindset that we have undertaken a targeted review of works related to the subsoil.

4.3 Materials and Methods

The general corpus we rely on includes written works, films, series, and video games. It was established based on searches for certain keywords (mainly from the IPCC[1] scenarios, which are considered the common base) in the titles and summaries of the works. We also included some classics or older works (more than 20 years old) that are less known but identified by other authors as containing interesting or precursor elements. We did not include myths and legends, such as the "myth of emergence," a family of stories that are older and universal, originating from Africa. The common narrative suggests that humans and animals once lived underground, but they emerged into daylight to spread across the globe (Le Quellec, 2022), implying that humans could feasibly return to live there.

[1] Intergovernmental Panel on Climate Change.

In the end, we selected nine works, either because the underground's role is predominant or original or because we considered the work as "representative" of "inevitable futures." For example, *The Road* by Cormac McCarthy (2006) is a particularly dark apocalyptic narrative. We also included an older work, the *Foundation* cycle by Isaac Asimov (starting in 1942), a reference in modern science fiction, which hypothesizes a possible evolution in the consideration of the underground in science fiction narratives. *Future Times Three* by René Barjavel caught our attention for its finale set in the year 100,000, offering a stunning vision of extreme Earth terraformation.[2] Another work holds a special place in our corpus due to its unique perspective on the relationship with the underground: *A Case of Conscience* by James Blish (1969). In this work, the underground is implicated in a disaster caused by the burial of our waste there, resulting in the generalized collapse of the Earth's crust and its subsequent submersion. In addition, we will sometimes reference other works not included in this restricted corpus when they offer relevant insights related to the underground. Ultimately, the corpus, as well as our analyses, presents a predominantly Western and contemporary view of the underground. We do not claim this corpus to be representative; rather, we consider it exploratory.

Simultaneously, we have occasionally consulted the website *L'Arbre des Possibles*,[3] launched in 2001 by French writer Bernard Werber. This collaborative site seemed to us a potential source for analyzing various content elements related to the theme of the underground.

[2] Mentioning terraforming in relation to the Earth 'sv environment may seem contradictory, as the concept is typically associated with making other planets or moons habitable for humans. However, there are circumstances where this principle applies to Earth, especially in projects focused on ecological rehabilitation or extensive environmental changes.

[3] "The tree of possibilities," not translated into English. https://www.arbredespossibles.com/base/scenarios.php [accessed on 27 February 2023].

List and Succinct Description of the Works

1. Isaac Asimov, "Foundation" cycle (1942–1986)
   Isaac Asimov's "Foundation" cycle, spanning from 1942 to 1986, narrates the implementation of the Seldon Plan in a distant future where humans have colonized the entire galaxy. Following the collapse of the first Galactic Empire, the Plan aims to establish a new Galactic Empire grounded on psychohistory, a science capable of predicting the historical evolution of large statistical masses. The series explores various planets, each with unique subterranean characteristics: Trantor, the Empire's capital, is built almost entirely underground with its surface covered in metal; Terminus, home to the Foundation, lacks extractive resources underground; and on Solaria, Spacers live entirely beneath the surface.
2. James Blish, "A Case of Conscience" (1969)
   The Earth is undergoing significant global warming, and pollution levels have become uncontrollable due to the accumulation of waste on the planet's surface, in the oceans, and underground. This situation has reached a point of no return, leading to the inevitable destruction of the planet.
3. Dmitry Glukhovsky, "Metro 2033" (2005)
   It is 2033, and the Earth's surface has become uninhabitable due to a nuclear accident, forcing Moscow's population to take refuge in the metro system. Survivor groups have formed small communities at different metro stations. Among these communities, with their varied forms of governance, conflicts frequently arise as they vie for domination or scramble for the last remaining resources.
4. Hugh Howey, "Wool" (2016)
   In a post-apocalyptic world where the Earth's surface is uninhabitable, a few thousand survivors live in a 144-story underground silo. For this micro-society to survive, everything is highly regulated, including births. Each floor serves a specific function, and everyone occupies a well-defined social and geographical position.[4]

[4] A comparison can easily be made with "Snowpiercer" (Lob et al., 1984), a French post-apocalyptic sci-fi comic that was adapted into a successful film in 2013 (Bong, 2013). After a climatic cataclysm, the surviving members of the human species are confined inside a massive train that travels endlessly. This train is intricately hierarchical, with front carriages for the aristocracy down to the tail carriages at the convoy 's rear, where the

5. N.K. Jemisin, "The Broken Earth" Series (2015–2017)
   In these books, orogenes, a type of human, can manipulate underground forces to create or quell earthquakes. Owing to this power, they are hunted, strictly controlled, and segregated from other humans. The Earth itself is a distinct character, with its climate undergoing dramatic changes throughout the story. Part of the narrative unfolds in an underground utopia where orogenes and non-orogenes live together in a community.
6. Cormac McCarthy, "The Road" (2006)
   In a cold, desolate post-apocalyptic world, a father and his son wander the roads. By chance, they discover underground shelter survivalists had created before the catastrophe. A place for rest and an oasis with a profusion of essential goods, it allows them to physically and mentally rejuvenate, living for a few hours as if in the pre-catastrophe world and regaining some dignity. However, the shelter's riches make them vulnerable, forcing them to stay only briefly.
7. Yannick Monget, "Gaïa" (2012)
   Overnight, the Earth transforms into an entity with agency, rebelling and subjecting humans to the treatment they have inflicted on plants and animals. An underground military bunker becomes a refuge offering all the comforts for autonomous living. However, the Earth manages to expel the protagonists from this sanctuary. Humans are harassed and pursued, and they find themselves no longer safe anywhere.
8. René Barjavel, "Future Times Three" (1944)
   A time traveler discovers the Earth in the year 100,000. "*Mountains have been leveled, oceans filled in, rivers buried, and lands flattened. An internal circulation has replaced the external water cycle: rain-river-sea-cloud-rain. Streams and rivers flow inside the globe in a perpetual motion maintained by the temperature differences of the underground. Man-made channels irrigate the meadows and orchards from below, provide the air with the moisture needed for life through plants, and carry the central fire's heat to the poles and the hemisphere threatened by winter. […] Everyone works for all, and all work for*

impoverished reside. Similar to "Wool," each carriage also serves a distinct function (army, nourishment, recreation), and the outside world is unreachable.

everyone, on or under a picturesque-free ground. [...] The new man does not practice agriculture per se. *He has simply exterminated all useless or harmful plants. He has also destroyed [...] all inhabitants of the waters, air, and land he had renounced using. Mammals have been reduced to two species: cows and pigs, which have become herbivores.*"[5]

9. "Fallout" Video Series (1997–2018).
   This video game series imagines an alternate history where the Cold War lasted much longer than it did in reality. During this period, the Vault-Tec corporation builds nuclear shelters, many of which are used to experiment with different societal profiles. A nuclear war eventually erupts, killing most of the inhabitants outside the shelters and turning the rest into monsters. In the twenty-second century (game No. 1), the hero, coming from the shelter that housed the "elites," is the first to emerge, tasked with finding repair materials in another shelter.

4.4 The Underground in Science Fiction: Between Invisibility and Utopia

Descriptive Aspects and Internal Operations of the Underground from Our Selection

Our initial hypothesis posits that underground areas can be classified based on various criteria related to their functions (Table 4.1):

- Their purposes: are they intended for resources, protection, or something else?
- Their physical attributes: shapes, locations (potentially extraterrestrial), depth?
- Their models of spatial organization: areolar, reticular, network-based?
- How is politics structured within these areas?
- What kinds of technology are connected to them? What degree of artificiality do they possess?

[5] Excerpt from René Barjavel 's book "Future Times Three" (1944).

Table 4.1 Overview of the descriptive aspects and internal workings of the studied works

Works	*Underground function*	*Shape, Location, Depth*	*Spatial organization patterns*	*Who built it?*	*Level/ Technological use*
Isaac Asimov, "Foundation" cycle (1942–1986)	Resource extraction and storage; shelter (radioactive exterior, space threats); optimized living space	Tanks (water on Trantor) For habitats: complex networks, more or less dense depending on the nature of the layers crossed	Networks	Inhabitants/ administration	Very important technology at the heart of the structures; robots (Earth and space planets)
James Blish, "A Case of Conscience" (1969)	Resource extraction and storage	Deep wells (>4,000 m) for waste storage		Manufacturer	Equivalent to today
Dmitry Glukhovsky, "Metro 2033" (2005)	Refuge (polluted outdoors)	Pockets linked by subway corridors	Networks	Administration, in the past	Same or lower than today
Hugh Howey, "Wool" (2016)	Refuge (polluted outdoors)	Silo	Bubble		Equivalent to today (mechanics and electronics)
N.K. Jemisin, "The Broken Earth" series (2015–2017)	Shelter (uninhabitable outdoors)	Far below surface, long and wide	Bubble	Pre-existing gallery + orogens	Superior technology
Cormac McCarthy, "The Road" (2006)	Shelter (dangerous outdoors)	Small underground shelter near the surface (bunker)	Bulle—Oasis	Private, in the past	No/little technology
Yannick Monget, "Gaïa" (2012)	Shelter (dangerous outdoors)	Multistory bunker deep underground	Bubble	Public, the army	Equivalent to today

(continued)

Table 4.1 (continued)

Works	*Underground function*	*Shape, Location, Depth*	*Spatial organization patterns*	*Who built it?*	*Level/ Technological use*
René Barjavel, "Future Times Three" (1944)	Global support and air conditioning	Temperature and humidity distribution channels	Networks	Shovel-men, in a human race that has become like a social insect	No technology; human body has adapted to its function
"Fallout" video series (1997–2018)	Shelter (dangerous outdoors)	Bunker	Bubble		Strong presence of technology/ cybernetics

Our initial analyses illuminate the broad spectrum of imagined situations, although underground spaces are often depicted as living spaces, predominantly of the bunker type.

The Underground as a Refuge

In most of the works, the underground serves as a refuge. This protective function arises either because the surface has become uninhabitable (*The Broken Earth* series, *Wool*) or because the outside is too polluted (*Metro 2033*, Glukhovsky, 2005) or dangerous (*The Road*, *Gaïa*, McCarthy, 2006, Monget, 2012; the *Fallout* (1997–2018) video series).[6] Historically marginalized in its status by humans and often depicted in Western mythologies as a disturbing place associated with death (Mattern, 2021), the underground thus becomes a habitable and protective replacement world. In some instances, such as the transport infrastructure in *Metro 2033* (Thibaud, 1995), the underground may have been adapted for use as a refuge before the need arose. From being a place of occasional passage, where it is not intended for long-term living, it becomes a place of permanent habitation. This shift is also linked to the established hierarchy between the ground and the underground; humans inhabit this lowest stratum, perceived as degraded. These margins, previously overlooked, are often associated with feelings of fear (Asmussen et al., 2020). Projecting oneself into these places often means a life without light, where the distinction between day and night becomes blurred, and where sight, such a vital sense outside, becomes secondary to hearing, as noted in works like *Gaïa* and *Metro 2033*. In our corpus, this alternative territorial choice provides a space to survive. Previously just a simple interstice, a territory of life for the excluded or people not wanted in plain sight, the underground is reinvested. Its welcoming function expands as it accommodates more people, here representing humans excluded from an uninhabitable world above. It is noteworthy that ultra-rich individuals are already implementing this function on an individual and reduced scale by creating comfortable underground bunkers. This action represents a form of secession from both the physical and social worlds, undertaken

[6] This protective function is found in many novels, including "The Man Who Lived Underground" by Wright (2003).

in anticipation of this future where the world is no longer inhabitable.[7] The notion of the protective bunker, an underground counterpart to the protective bubble, is extensively explored in works like *Wool, The Road,* and *Fallout*, among others, and will be discussed further. Versions of underground life in a network are also imagined (e.g., *Metro 2033*). While the underground serves as a space of movement between various places, there seems to be no link between the models of spatial organization and the type of political organization governing life underground, nor with the use of advanced technology. However, we will see in the following section how political governance is closely tied to perceptions of the surface/outside world, despite the underground's vulnerability to climate change (Rotta-Loria, 2023).

The Underground as a Source of Knowledge

The underground also acts as a repository of truth and/or knowledge about the future: it is here that one can find traces of the past, such as the archives in *Foundation*, or even of ancient civilizations (*The Broken Earth* series). In this sense, the underground is a refuge not only for humans but also for the "treasures," whatever they may be, that have been concealed or buried there. The characters in the books mine the knowledge preserved in these undergrounds by previous generations and use it to understand and untangle the plot. It is there that secrets are revealed, providing the necessary answers for the story to progress. Thus, the underground serves as a past that helps piece together elements to build the future.

An Underground Where the Biodiversity Is Absent or Hostile

While it is a refuge for humans and their buried treasures, the underground is generally perceived as a lifeless space—a blank slate that can be colonized, reshaped, or even destroyed. This concept is exemplified in *Future Times Three*, where the underground's biodiversity (as well as that of the entire Earth) has been completely annihilated to create a perfectly sterilized hydroponic environment. It is interesting to note that when

[7] https://www.forbes.fr/lifestyle/quand-les-riches-se-preparent-au-pire-dans-des-bunkers-de-luxe/ [accessed on 10 September 2023].

underground environments are inhabited by living beings, they are typically portrayed through their struggles to protect their territory from human intruders. *The Broken Earth* series mentions a few arthropods whose actions lead to the disappearance of the city beneath the soil. In *Gaïa*, a particularly hostile version of biodiversity is depicted through a root system that pierces the bunker, precipitating the protagonists' escape. Describing ways of living with this biodiversity and preserving it exposes a lack of vocabulary (Steward, 2023). Apart from these examples, signs of the presence of living species are very subtle and reveal our societies' lack of knowledge about these environments, despite laudable and ambitious attempts to promote an "underground culture."[8] Thus, inhabiting, living, and exploiting the underground raises questions about relationships with the species that live there, the "inhabiting organisms" that are rarely highlighted in the works studied. Their potential as a resource enabling life underground is also not addressed.

The Underground as a Resource, a Theme Varying According to Epochs and Media

Within the presented corpus, *A Case of Conscience* by James Blish and the works of Asimov's *Foundation* cycle occupy a unique position. These two works are notable for being the oldest in the corpus: 1942 to 1953 for the first *Foundation* trilogy and 1982 and 1986 for the continuations; and 1969 for *A Case of Conscience*. They also concretely address undergrounds in terms of resources, both in raw materials and as storage areas, in connection with their own geological characteristics. Asimov explored a comprehensive set of modalities for using and occupying the underground. Regarding the underground as a source of materials, he mentions the absence (initially problematic then ultimately beneficial) of mining resources on Terminus, the varying compositions of the underground in the tunnels of Solaria, or the artificial groundwater storage on Trantor. The *Foundation* cycle encompasses all the possibilities offered by undergrounds as described above: undergrounds of different planets (or their satellites) function as refuges from a radioactive environment (Earth), lack of atmosphere (Moon), hostile conditions (Solaria), as a place for resource optimization (Trantor), or as a model for a new society (Solaria); as a place

[8] For example, "Terra Forma: manuel de cartographies potentielles" by Alexandra Arènes, Axelle Grégoire, and Frédérique Aït-Touati (2019, Éditions B42).

where the Earth's archives are (literally) found (many planets explored by the heroes) or the "treasure" that enables a desirable resolution to the story (the character of Fallom). James Blish, on his part, directly addresses the resources of the underground by describing deep drilling wells and geological mechanisms that lead to major earthquakes. At the same time, Franck Herbert, in *Dune* (Herbert, 1965), bases part of his novel on an essential resource, the spice, which is the result of the maturation of the excretions of sand trouts (the primary stage of the sandworms) in the presence of water and carbon dioxide.[9]

A Perception of Undergrounds That Changes Over Time

These examples suggest a distinct chronology in how the underground is portrayed in works of science fiction. The concept of the underground as a resource was prominent until the 1970s but subsequently seems to fade. To reinforce this important observation, based on our limited corpus, we explored other types of science fiction narratives.

In 2001, science fiction writer Bernard Werber, known for numerous science fiction books, launched the website *L'Arbre des Possibles.*[10] This project reflects the public's diminishing interest in this theme over the past two decades. Out of 9,257 scenarios imagined (as of 2022), the term "underground" (or "ground") appears in only 68 scenarios (or 0.73% of the total). Of the 69 occurrences noted (one scenario presents two different visions of the underground), the undergrounds are present 43.5% of the time as shelters—often bunkers—or protections against the surface of the planet that has become uninhabitable. Only eight scenarios (11.6%) consider the underground in its dimension as a resource (e.g., water or energy). This additional analysis corroborates the idea that undergrounds are rarely featured in current science fiction texts, and when they are, they are primarily portrayed as shelters.

[9] The process of creating spice is reminiscent of that of oil. https://www.larecherche.fr/l%C3%A9pice-de-%C2%AB-dune-%C2%BB-bient%C3%B4t-disponible-sur-terre [accessed 27 February 2023].

[10] Not translated into English. https://www.arbredespossibles.com/base/scenarios.php [accessed 27 February 2023].

Video Games, Deeply Rooted in Undergrounds

Video games represent a contemporary domain where science fiction undergrounds are well represented. Several globally popular video games incorporate the concept of terraforming,[11] whether set in science fiction universes or not: *Minecraft* (Mojang Studios, 2011), *Rimworld* (Ludeon Studios, 2013), *Craft the World* (Dekovir Entertainment, 2014), and *Oxygen Not Included* (Klei Entertainment, 2019) for example. In these games, the concrete dimension of the underground is pronounced. To survive, players must extract resources, creating underground voids where they can/must construct shelters. Thus, in these games, the underground is a resource both through the materials it provides and as a form of shelter. The game, which goes through a mapping of places, creates a continuity between the interior and the exterior, bubbles, and network, and it imposes the materiality of the underground on players. This contrasts with cinema and literature, which can more easily create spatial and temporal ellipses to advance the narrative. In video games, at least in management or sandbox games, players are encouraged to create their own narratives as they see fit, even if it means frequent revisions to improve their strategy.

Thus, our limited corpus reveals that recent literary science fiction tends to erase the materiality of the underground, treating it as if it were an ancillary, unimportant element. This leads to the paradox that while for some authors the future lies in the underground, it is often ignored.

4.5 Living in the Underground: Imagining Tomorrow's Governance?

Beyond the descriptive elements gathered in this initial table, we have taken an interest in other, less obvious criteria derived from the dominant models of subterranean spaces—namely, shelters/bunkers that protect the population from a hostile exterior.

Indeed, these inhabited undergrounds, serving as refuges for the population, are as much enclaves combining isolation and security as islands

[11] Terraforming is a game mode in which players can modify or reshape the environment of a planet or virtual world to make it habitable, productive, or simply in line with their preferences. Terraforming is often associated with simulation, management, construction, or strategy games [https://fr.wikipedia.org/wiki/Terraformation, accessed 10 September 2023].

laying the foundation for a utopia, that is, the potential realization of an ideal society confined to a limited space.[12] As Jameson (2005) notes, "*The utopian space constitutes an imaginary enclave within the real social space, or in other terms, [...] the very possibility of utopian space results from a spatial and social differentiation.*" A bunker, a silo, or an underground shelter provides the perfect setting to experiment with various models of utopian societies.

Thus, our corpus reveals a great diversity of imagined modes of governance, which is central to today's environmental transition issues. What do these works show us about how science fiction envisions these modes of governance?

Our corpus includes systems of communities that are more or less dependent on one another, featuring democratic or military systems and varying levels of authoritarianism and/or hierarchy. In *Foundation*, an empire governs the universe, while in *Metro 2033*, small communities operate under Soviet-style systems. *Gaïa* presents a professional military organization where everything is meticulously ordered and organized. By contrast, in *The Road*, the political system has collapsed, leaving only a family unit and ruins of the old world. *The Broken Earth* series, for a few chapters, imagines a utopian society where everyone is welcomed with their differences, treated equally, and their needs considered. Access to the underground is granted only once this "social contract" is accepted.

We propose to examine a few specific aspects of governance systems, such as the regulation of exchanges between subterranean spaces, control of access, and dependency on technology.

The Porosity of the Subsoil and the Issue of Control

In these underground shelters, the issue of porosity seems central as it determines the vulnerability to external elements, whether they be climatic, related to biodiversity, or human conflicts. Porosity appears to be inversely proportional to the level of danger experienced. Hermetically sealed subsoils, like that in *Wool*, are almost completely protected from the dangers of adjacent spaces (with the exception of the deepest levels),

[12] The parallel model on the surface is the "capsule" or protective bubble (or dome), commonly seen in science-fiction literature and films. Apart from the aerial aspect, it is a model frequently used in mining research, creating extraterritorial enclaves isolated from local contexts (Donner, 2011; Joncheray, 2020).

whereas those in *Metro 2033*, which feature numerous underground and aerial openings, are exposed to all dangers... except when nature itself intervenes, as in *Gaïa*.

Porosity reduces control over all facets of society, thus linking it to greater individual or even collective freedom, as in *Metro 2033*. However, increased porosity, synonymous with greater dangers, leads to increased militarization or combat training and the necessity to control access by various means (armies, technologies).

The second notable characteristic, generally related to the political systems implemented, is the possibility (or lack thereof) to exit the subsoil. The more hermetic it is, the more the constraints of survival in a closed environment seem to impose a control policy and, thus, strong or even authoritarian governance. *Wool*, which is hermetically sealed, appears democratic but is totally controlled by a higher authority; the bunkers, highly controlled but with possible exits, also enforce strong control policies (*Gaïa, Fallout*), whereas *Metro 2033* is much more anarchic overall.

Technology Versus Nature

These two aspects underscore the role of technology: the tighter the closure, the greater the need for self-sufficiency, which in turn leads to the deployment of advanced technologies (*Gaïa, Fallout, Foundation*). The question of self-sufficiency also brings up the role of nature, especially nurturing nature. Undergrounds are depicted as an excellent medium for exploring the absence of nature. As analyzed by Williams (2008), the underground can be viewed as "*a visionary site for technological futures*." She particularly notes that a defining aspect of these futures is "*the exclusion of nature—of biological diversity, seasons, plants, the sun, and the stars*," a sort of underground laboratory that "*pushes to the extreme the ecological simplification of modern cities where it sometimes seems that humans, rats, insects, and microbes are the only remaining formsof wildlife*" (Williams, 2008). Starting from the premise that the underground "*appears to be the apotheosis of this very modern separation between nature and culture*" (Hawkins, 2020), its portrayal here facilitates the projection of a world without nature.

However, in the two works set very close to our time immediately post-disaster (*The Road, Metro 2033*), technology is notably absent. Instead, there is a level of technology less advanced than today, hindered by the impossibility of maintaining existing facilities and machines. Outside the underground, nature begins to reclaim its place. One of the challenges for those living underground will therefore be to combat this expansion of nature (*Metro 2033*) and to implement new technologies suited for survival underground and for eventually reclaiming the surface. It suggests that our future might be conceived as a technological race against nature. Yet, the desire to make the most of one's environment does not always stem from an intensive use of technology (*Future Times Three*). The extreme modification of the Earth's environment often results from establishing a strange but perfectly functional symbiosis: the outcome of utilitarianism with an omnipresent "nature" yet of absolute scarcity, including underground.

Returning to the Surface and the Myth of the Lost Paradise

We also explored how the protagonists view the underground and their relationships with the ground/above/outside. Generally, a distinction is made between works where the underground is the last habitable place (*Wool*, *Metro 2033*) and those where there is opposition, or even confrontation, between those underground and those who have remained on the surface. In this case, the underground is portrayed either as the place to be (*The Broken Earth* series, *Gaïa*) or as a degraded place. Generally, with the notable exception of Asimov's works where the underground (of Trantor, Solaria, or the Earth) has become the new norm (but here, the hero finds this norm undesirable), the outside remains the reference point, whether it is accessible or merely fantasized about, for what life on the planet *should* be like. This is particularly evident in *The Road*, where the abundance of goods in the bunker contrasts sharply with the outside world. The habits and objects mentioned resemble relics from the past, whose habits must be relearned (e.g., making hot chocolate), with the father lamenting his inability to share this world with his son.

Thus, the last column of Table 4.2, which categorizes the underground as either merely a support (in the majority of works) or as an actor (expressed in anthropomorphic terms like submission, revolt, or revenge), can be seen as reflecting a sense of guilt borne by the human species for the planet's destruction. In *Gaïa* and *The Broken Earth* series, the avenging Earth comes to punish humans for having tampered too much with the heart of the planet, for having altered it irreversibly, reminding

Table 4.2 Synthesis on surface/outdoor relationships, perception of the protagonists, and potential role played by the underground

Works	*Governance*	*Exit to the outside/ surface*	*Self-sufficiency*	*Porosity*	*How the protagonists view the subsoil*	*Relationship of basement to ground/ overground/ outside*	*Underground as a player or just as support?*
Isaac Asimov, "Foundation" cycle (1942–1986)	Miscellaneous	Trantor or Solaria: we rarely go out. Free, but not loved	No	Low	Normality	Disinterest in the surface (but negative judgment of the hero)	Support
Dmitry Glukhovsky, "Metro 2033" (2005)	Independent communities, various systems	Possible/ free—but controlled	No	High = danger	Fear -	Everyone underground; dreams of returning to the surface	Support
Hugh Howey, "Wool" (2016)	Hierarchical authoritarian democratic system	Impossible	Yes	Virtually nonexistent; maximum safety	Household +	Everyone underground; dreams of returning to the surface	Support
N.K. Jemisin, "The Broken Earth" series (2015–2017)	Democratic	Possible/free, but controlled	Yes	Low	Refuge (positive image, unknown technology, but value +)	In the basement: the privileged	Revenge

(continued)

Table 4.2 (continued)

Works	*Governance*	*Exit to the outside/surface*	*Self-sufficiency*	*Porosity*	*How the protagonists view the subsoil*	*Relationship of basement to ground/overground/outside*	*Underground as a player or just as support?*
Cormac McCarthy, "The Road" (2006)	No	Possible/free, but controlled	No	Low	Refuge (but seen as confinement: -)	Underground: the lucky ones who fall on it	Support
Yannick Monget, "Gaïa" (2012)	Military	Possible	Yes	Strong	Refuge and trap	In the basement: the people who managed to survive	Revenge
René Barjavel, "Future Times Three" (1944)	Social insect self-organization	Not applicable; a functional place for everyone	Yes	Nonexistent; to each their own functional place	Neutral, no representation	Instinctive collaboration	Support
"Fallout" video series (1997–2018)	Miscellaneous	Possible/free, but controlled	Yes	Low	Refuge—neutral	Privileged underground: those who survived on the surface turned into monsters	Support

them that they are often the cause of the evils forcing them to seek refuge underground.

4.6 Conclusion

By using forms of amplification and exaggeration, science fiction serves as a powerful revealer of contemporary concerns and anxieties. From this perspective, it is clear that the underground realm does not occupy a central position in science fiction narratives. When addressed, it is often in a utilitarian context, highlighting its role in supporting human survival. This perspective almost directly contrasts with the very idea of nature, as if the underground were not a part of it. Generally, the science fiction narratives in our corpus tend to explore only three options for humanity's survival: an increased dependency on technology, the relentless pursuit of progress (understood as the ability to dominate nature), or the abandonment of our organic connection to the planet.

However, while it is recognized that science fiction "reflects" the societies from which its authors emerge, it should not be concluded that fiction merely replicates our society. It also helps shape it. In this regard, the connection between fiction and the future of underground spaces (especially their management and geotechnical exploitation) takes on another dimension.

For Chelebourg (2012), examining fiction places one in an unusual position: It is both a "recording" site of society and a place where society is constructed. In other words, fiction influences people's mindsets, opinions, ideas, or what we might call "the spirit of the times." The transmission, or even creation, of norms are indeed traditional functions of art and of fiction in particular (Picholle, 2012). "*Through a deconstruction of certainties and an exploration of possibilities, but also of impossibilities, science fiction contributes to the transformation of social representations from which we act and explainthe world*" (Girard and Gendron, 2010), as demonstrated by the study of Coruscant in *Star Wars*, which has become a key to understanding the future of today's cities (Musset, 2005). Meanwhile, Laugier (2019) explores how the heroes of the *Star Wars* saga have evolved into a contemporary myth that inspires our vocabulary and morality.

While some question the formative role of fiction, Picholle (2012) reminds us that "*science fiction has already proven itself in this matter, particularly in the space context,*" as NASA itself has acknowledged. Space exploration became normalized banal and politically acceptable through the numerous articles, books, and films dedicated to it before it even occurred, "*the engineers of the 1960s being the teenagers of the1950s.*" Similarly, McCurdy (1998) mentions the influence of the *Star Trek* series and numerous science fiction narratives on the launch of the Apollo mission. According to Michaud (2020), science fiction "*is instrumentalized by aerospace engineering communities and almost serves as managerial literature in this field.*" As a result, science fiction courses are provided to NASA engineers. This inspiration also led to the ESA's ITSF (Innovative Technologies from Science Fiction) project in Europe, designed to bridge science fiction with research and development projects.

The way we think about the underground—and our future—might benefit from reconsidering our past. For example, Guillaume (2019) suggests that a new regime of historicity is emerging, where the past is not seen as archaic but full of possibilities: "*Communities to reconnect with the earth, to feel terrestrial again, must reconnect with a revisited past. […] Thus, to overcome presentism, this stance invites linking the past to the future to shed old-fashioned progressivism. […] The modernization project wrongly discredits the past because it emancipates without recognizing our dependencies. It disconnects fromthe earth, from material conditions, the dreams of emancipation. […] The concept of retrofuturism reconciles thetwo temporalities, past and future. It rejects both backward flight and forward flight.*"

To think about our relationship with the underground in time scales and volumes beyond our immediate experience, we should develop new imaginaries to "fill the void" and possibly harness it. "Underground" geoengineering is highly advanced, with various sites for capturing greenhouse gases being developed and tested, as well as the burying of (nuclear or other) waste. However, these aspects are not prominently featured in mainstream works of science fiction, unlike "atmospheric" or "surface" geoengineering such as climate regulation via satellite networks, dispersion of substances in the atmosphere or oceans, or altering ocean currents (Deletraz and Rebotier, 2022).

On this basis, however, the underground could yet become a locus for utopian dreams. For Kearnes and Rickards (2017), in the context of an uncertain environmental future, the underground could indeed be seen as an earthly savior by presenting the burial of radioactive and carbon waste as a long-term solution to the problematic by-products of our industries and lifestyles. These burials and other sequestrations could then be a mirror image of the mining, drilling, and hydraulic fracturing processes. Scientists, policymakers, and organizations might work to prioritize certain cultural productions over others to reinvent a more global underground imaginary, conducive to its intensive exploitation, in the same way that Michaud (2020) describes the strategic interest in fostering a culture "stimulating the imagination" and favorable to innovation in the European space sector.

By framing underground geoengineering initiatives within a broader context of great mythological narratives of planetary balance and viewing the Earth as a superorganism (Gaïa) capable of regenerating all its elements, such a conceptual framework could foster a culture and environment that attract investors and policies (public or private) involved in the development of certain technological aspects of ecological transitions requiring profound societal choices.

The central question then becomes identifying which imaginative narratives, among those already explored in existing works or under development, will resonate with the concerns and aspirations of societies. Which will be promoted or possibly censored by certain interest groups? Which will influence the collective imagination of our technoscientific societies enough to construct performative representations of our future? These considerations have had and will continue to have a significant impact on the nature of debates guiding the concrete choices that are or must be made.

Revisiting Groves's (2017) analysis, we understand the concern about the potential instrumentalization of science fiction within the context of the unequal anticipation capabilities described by the author. Science fiction could become an additional tool in a vast arsenal to impose or suggest certain visions of the future. However, imagine if the style of anticipation that Groves advocates for were to develop—anticipation based on lived and affectively engaged futures, focusing on experiences, emotions, and attachments. In this scenario, science fiction narratives could become catalysts for somewhat rebalancing the unequal distribution of anticipation capabilities.

These perspectives highlight the importance of an ethical approach to the future based on collective care, responsibility toward future generations, and interdependence among humans, and between humans and non-humans, between biotic and abiotic elements, rather than on pursuing short-sighted present interests (Adam and Groves, 2011). It also involves recognizing that soils and subsoils need to be protected and preserved starting now.

Acknowledgements We wish to express our gratitude to the editorial committee for its comments on the initial drafts, its reading suggestions, and for the enhancement recommendations.

References

Adam, B. (2010). History of the future: Paradoxes and challenges. *Rethinking History, 14*(3), 361–378.

Adam, B. (2023). Futures Imperfect: A Reflection on Challenges. *Sociology: The Journal of the British Sociological Association*, *57*(2), 279–287.

Adam, B., & Groves, C. (2011). Futures tended: Care and future-oriented responsibility. *Bulletin of Science, Technology & Society, 31*(1), 17–27.

Asmussen, T., Berger, S., Elsig, A., & Hoenig, B. (2020). Sous le sol: Editorial. *Traverse. Revue d'Histoire, 2*, 17–25.

Barreau, B., Ducos, G., & Teillant, A. (2012). La science-fiction, du miroir de nos sociétés à la réflexion prospective. *La note d'analyse – Développement durable, Centre d'analyse stratégique*, (311), 11.

Beck, U. (2001). *La société du risque. Sur la voie d'une autre modernité* (L. Bernardi, Trans.). Aubier.

Boutaud, A., & Gondran, N. (2020). *Les limites planétaires*. La Découverte.

Chelebourg, C. (2012). *Les Ecofictions: Mythologies de la fin du monde*. Les Impressions Nouvelles, collection "Réflexions faites." ISBN: 9782874491405.

Clayton, S., Manning, C. M., Krygsman, K., & Speiser, M. (2017). *Mental health and our changing climate: Impacts, implications, and guidance*. American Psychological Association, and ecoAmerica.

Deletraz, G., & Rebotier, J. (2022). *Explorer notre futur climatique à travers la science-fiction*. Conférence, Unight – Nuit Européenne des Chercheurs, Arudy, France.

Domenach, J.-M. (1966). Le bon usage de l'avenir. *Esprit, 346*(2), 271–281.

Donner, N. (2011). Notes sur la dimension immunitaire des enclaves pétrolières. *Echogéo, 17*.

Dussert, M. (2020). La « Red Team », cellule de science-fiction de l'armée française, imagine les pires dystopies. *L'ADN*. Retrieved from https://www.ladn.eu/mondes-creatifs/la-red-team-nouvelle-cellule-de-science-fiction-de-larmee-francaise-verse-dans-la-dystopie

Fortey, R. (1994). *The hidden landscape: A journey into the geological past*.

Gendron, C., Audet, R., & Girard, B. (2012). Thinking post-ecological societies with science-fiction. *Les Cahiers de la CRSDD* – Coll. Recherche, (9).

Gimbert, L. (2004). La prospective: Pour un retour aux fondamentaux. *Management & Avenir, 2*, 83–95. https://doi.org/10.3917/mav.002.0083

Girard, B., & Gendron, C. (2010). Quand l'imaginaire produit du social. *Les Cahiers de la CRSDD* – Collection La science-fiction: un regard sur l'organisation des sociétés postécologiques, (3).
Groves, C. (2017). Emptying the future: On the environmental politics of anticipation. *Futures, 92*, 29–38. https://doi.org/10.1016/j.futures.2016.06.003
Guillaume, C. (2019). Le futur ancien n'est pas notre avenir. *Philosophique*, *22*. Retrieved from http://journals.openedition.org/philosophique/1279
Hartog, F. (2003). *Régimes d'historicité. Présentisme et expérience du temps*. Le Seuil.
Hawkins, H. (2020). Underground Imaginations, Environmental Crisis and Subterranean Cultural Geographies. *Cultural Geographies, 27*(1), 3–22.
Henshel, R. L. (1981). Insights from the history: Evolution of controversial fields: Lesson from the past to futures. *Futures, 13*(5), 401–412.
Jameson, F. (2005). *Archéologies du futur, T1 Le désir nommé utopie*. Max Milo.
Jonas, H. (1995). *Le Principe responsabilité*. Champs Flammarion.
Jonas, H. (1997). *Pour une éthique du futur*. Rivages.
Joncheray, M. (2020). Espoirs et désillusions du développement minier en contexte post-conflit. Un Exemple Congolais. *Mondes En Développement, 189*(1), 31–53.
Jouvenel (de), H. (2016). *Science-fiction et prospective*. Futuribles, (413).
Kearnes, M., & Rickards, L. (2017). Earthly graves for environmental futures: Techno-burial practices. *Futures, 92*, 48–58.
Laugier, S. (2019). *Star Wars, la force d'un mythe contemporain*. Le monde des religions, Hors-serie 32.
Le Quellec, J. (2022). VII. L'émergence des animaux et de l'humanité. In: J. Le Quellec, *La caverne originelle: Art, mythes et premières humanités* (pp. 617–657). La Découverte.
Lessault, B. (2004). F. Hartog. Régimes d'historicité. Présentisme et expérience du temps. *L'orientation scolaire et professionnelle, 33*(3), 479–483.
Mattern, J. (2021). *Les Titans, les sous-sols et la transition - Extractivisme minier et logiques de déferlement*. Colloque Sols et sous-sols dans la transition socio-écologique - Enquêtes, récits et imaginaires des sols et sous-sols à l'heure des transitions, 10–11 juin 2021, Grenoble.
McCurdy, H. E. (1998). *Space and the American imagination*, Smithsonian Books.
Meheust, B. (2012). La pente « biocidaire » du monde moderne. In *La nostalgie de l'Occupation: Peut-on encore se rebeller contre les nouvelles formes d'asservissement* ? La Découverte.
Michaud, T. (2020). *Le projet spatial européen, entre pragmatisme et imagination*. L'Harmattan.

Missaghieh-Poncet, J. (2023). Aux frontières de la géothermie profonde: l'émergence du milieu « sous-sol » dans un contexte de transition énergétique. *Géographie*. Université de Pau et des Pays de l'Adour, X. Arnault de Sartre (ed.).

Montgomery, D. (2015). *The hidden half of nature: The microbial roots of life and health.*

Morin, E. (2020). *Vers l'abîme ? 10 essais pour penser l'avenir* (p. 9). Flammarion.

Musset, A. (2005). *De New York à Coruscant. Essai de géofiction.* P.U.F.

Owens, S. (2004). Siting, sustainable development and social priorities. *Journal of Risk Research, 7*(2), 101–114.

Picholle, E. (2012). Penser l'impensable: terra-former une autre planète pour préserver la nôtre ? In *De la science-fiction à la réalité. Actes de colloque - Centre d'analyse stratégique.*

Polère, C. (2012). *La prospective - Questions actuelles.* Grand Lyon/Direction de la Prospective et du Dialogue Public.

Rotta-Loria, A. F. (2023). The silent impact of underground climate change on civil infrastructure. *Commun Eng., 2*, 44.

Rumpala, Y. (2010). Ce que la science-fiction pourrait apporter à la pensée politique. *Raisons politiques, 4/2010* (*40*), 97–113.

Rumpala, Y. (2016). Science-fiction, Spéculations écologiques et éthique du futur. *Revue Française D'éthique Appliquée, 2*(2), 74–89.

Rumpala, Y. (2018). *Hors des décombres du monde, Écologie, science-fiction et éthique du futur.* Champ Vallon.

Scouarnec, A. (2008). Plaidoyer pour un « renouveau » de la prospective. *Management & Avenir*, *2008/5* (*19*), 171–186.

Smida, A. (2004). *La prospective: entre ambitions et pratiques.* Les docs d'Aleph, Commissariat Général du Plan, 10.

Steffen, W., Richardson, K., Rockström, J., Cornell, S. E., Fetzer, I., Bennett, E. M., Biggs, R., Carpenter, S. R., de Vries, W., de Wit, C. A., et al. (2015). Planetary boundaries: Guiding human development on a changing planet. *Science.*

Steward, K. (2023). *Le Futur au pluriel: réparer la science-fiction, Inframonde.*

Thibaud, J.-P. (1995). Visions du public en sous-sol. EUS 95. *Espace et urbanisme souterrains: Actes de la 6ème conférence internationale*, Paris, 26–29 septembre 1995.

Vidalenc, E., Combet, E., Bergey, J.-L., Quiniou, V., & Marchal, D. (2022). Quatre scénarios pour la transition écologique - L'exercice de prospective de l'ADEME Transition(s) 2050. *Futuribles, 3*(448), 5–21.

Williams, R. (2008). *Notes on the Underground: An Essay on Technology, Society, and the Imagination.* MIT Press.

Works cited

Books and comics

Asimov, I. (2009). *Fondation* (intégrale) [Foundation]. Gallimard. (Original work published 1942–1986)

Barjavel, R. (1996). *Future Times Three*. Gallimard. (Original work published 1944)

Blish, J. (2014). *Nous mourons nus* [A Case of Conscience]. Passager Clandestin. (Original work published 1969).

Glukhovsky, D. (2005). *Métro 2033* [Метро 2033]. L'Atalante.

Herbert, F. (2012). *Dune*. Pocket Science-fiction. (Original work published 1965)

Howey, H. (2016). *Silo* [Wool]. Le Livre de poche

Jemisin, N.K. (2017). *Les livres de la terre fracturée* (intégrale) [The Broken Earth]. Orbit. (Original work published 2015–2017)

Lob, J., Rochette, J.-M., Legrand, B. (2014). *Transperceneige* (intégrale) [Snowpiercer]. Casterman. (Original work published 1984)

McCarthy, C. (2006). *La Route* [The Road]. Éditions de l'Olivier.

Monget, Y. (2012). *Gaïa*. Bragelonne.

Wright, R. (2003). *L'homme qui vivait sous terre [The Man Who Lived Underground]*. Gallimard.

Movies and series

Bong, J.-H. (Director). (2013). *Snowpiercer*. Moho Films

Lucas, G. (Creator). (1977–2019). *Star Wars*. LucasFilm.

Fallout [Video series]. (1997–2018). Black Isle Studios/Bethesda Softworks.

Star Trek [Television series]. (1966–1969). Paramount Television Studios.

Video games

Dekovir Entertainment. (2014). *Craft the world*.

Klei Entertainment. (2019). *Oxygen not included*.

Ludeon Studios. (2013). *Rimworld*.

Mojang Studios. (2011). *Minecraft*.

CHAPTER 5

Perspectives on Soil Depth: Jacek Yerka's SF Architectures

Julie Beauté

List of Abbreviation

SF Science fiction, scientific fact, and speculative fabulation (following Donna Haraway)

5.1 Introduction

Jacek Yerka, a Polish artist born in 1952, uses acrylics to paint disturbing worlds populated by strange creatures and paradoxical infrastructures. His landscapes feature disconcerting arrangements and sometimes incongruous angles of view, and they blur orientation and spatial reference points. By combining reality and imagination with striking colors, the artist puts spatial dimensions in tension and challenges our conception of the ground and the underground: He subverts the codes of perspective and deconstructs the traditional vertical representation of the layers

J. Beauté (✉)
ENS Ulm, PSL, Paris, France
e-mail: julie.beaute@ens.psl.eu

© The Author(s), under exclusive license to Springer Nature Switzerland AG 2025

O. Labussière et al. (eds.), *Back to the Ground*,
https://doi.org/10.1007/978-3-031-88888-5_5

of earth. He makes thickness and depth tangible, as in his emblematic painting, *Tectonics*: Beneath a house, the ground layers converge to form a vast ship flowing peacefully along a lava river (Jacek Yerka, *Tectonics*, this painting can be viewed online at: https://www.wikiart.org/en/jacek-yerka/tectonics). How does Yerka move beyond viewing the soil in terms of a vertical stack of smooth, horizontal strata to explore its depths? In this paper, I try to explain the transition from univocal verticality to multi-dimensional depths through his fantastical architecture. The paintings I discuss are accessible online on the visual arts encyclopedia WikiArt, which you can access by using QR codes for a playful immersion into the images. I aim to show how the artist invites us to dig into our imaginations, pierce the watertight surfaces of modernity in search of connections, and break free from the shackles from below. This exploration of the work of Jacek Yerka sketches some paths toward an ecology of depths.[1]

5.2 From Magic Realism to SF Architecture

The pictorial worlds of Jacek Yerka's paintings (Cowan, 1994), though suffused with a surrealist atmosphere, seem to be associated with magic realism. Introduced in 1925 by German art critic Franz Roh to account in painting for elements perceived as supernatural or irrational in a realistic setting (Roh, 1925) before being taken up by Latin American avant-garde literature, magic realism presents a recognizable, even familiar, reality that becomes the site of paranormal and dreamlike manifestations. Unlike classical fantasy, the emergence of the irrational or the imaginary does not present as a starting conflict between the rational and the supernatural. Within this singular perception of reality, magic realism aims to unify opposing categories: rational and irrational, life and dreams, reality and imaginary (Denis, 2002). Yerka fits well into this approach by going beyond these dichotomies and proposing a renewed apprehension of the world (Ellison & Yerka, 2006). His *Epitaph*, for example, depicts a seemingly realistic barn (Jacek Yerka, *Epitaph*, this painting can be viewed online at: https://www.wikiart.org/en/jacek-yerka/epitaph).

[1] The notion of "ecology" must be understood here in a double sense: On the one hand, it involves considering the thickness of the soil from an environmental perspective, in contrast to an extractivist approach. On the other hand, it aims to highlight the issue of depth through a relational lens by emphasizing the networks of relationships that shape soils and subsoils.

Upon closer inspection, however, the viewer discovers the thatched roof is a meadow, the main room, filled with strange machines, opens onto an idyllic fountain, and the side doors lead to an enchanted forest and a cosmic sky. The painting echoes an earlier one, *Philip's Site*, as if the artist were paying tribute to a departed loved one by linking life and death. Yerka's paintings, primarily made with acrylics, plunge us into dreamlike visions inspired by his dreams, travels, and childhood memories. They also draw heavily on fifteenth-century Flemish artists, particularly from the so-called Flemish Primitive movement, such as Hieronymus Bosch and Jan van Eyck. By combining technical precision with unlimited imagination, Yerka disrupts pictorial arrangements and spatial dimensions.

The artist's compositions leverage the mixtures and ambiguities of magic realism to depict architectures that could be described as SF architectures—in Donna Haraway's sense of science fiction, scientific fact, and speculative fabulation (Haraway, 2016). The buildings and infrastructures of these paintings indeed serve as sites for the intermingling of concrete and factual details, futuristic elements, chimerical perspectives, and dream-like atmospheres. Made possible by drawing, yet transcending the status of a simple sketch, Yerka's SF architectures are composed of attachments and detachments, cuts and knots, and consequences without determinism. In the painting *Cowan City*, for example, Yerka distinguishes several levels without discontinuity: A city is nestled in a rocky mound, which turns out to be a tree, rooted in the clouds thanks to a root pipe (Jacek Yerka, *Epitaph*, this painting can be viewed online at: https://www.wikiart.org/en/jacek-yerka/epitaph). Similarly, in the painting *Loading Cities*, the city sharply contrasts with a rural landscape while inscribing itself in it; it floats in the sky, ambiguously either assembling or dissipating like the smoke from a volcano (Jacek Yerka, *Cowan City*, this painting can be viewed online at: https://www.wikiart.org/en/jacek-yerka/cowan-city). All in all, the architectures stem from an approach that is both empirical and speculative. It is well understood by the painter, who constantly demonstrates that the factual and the fabricated call to and support each other. Yerka offers architecture a sidestep, in the sense that his fantastical universe challenges the role of imagination in the thinking and practice of architecture.

The Polish painter indeed captures space through imagination, resonating with the concept of topophilia that Gaston Bachelard discusses (Bachelard, 2012). The SF architectures, like the Bachelardian house, serve as a motif to apprehend the relationship to space, in both its reality and its virtuality. Yerka evokes images of retreat and openness, stirring up dreams and memories and invoking both shelter and metamorphosis in the storm. However, Yerka questions the importance given by Bachelard to verticality, from cellar to attic, which endowed the house with its cosmicity. For example, Yerka's "4siders," like the painting *Boudoir* (Jacek Yerka, *Loading Cities*, this painting can be viewed online at: https://www.wikiart.org/en/jacek-yerka/loading-cities), represent interiors inspired by his grandmother's house, but in such a way that the painting can be viewed from all angles: It has no definitive top or bottom. This creates an atmosphere of familiarity in the room that is explicitly not depicted through a vertical axis.

By contrast, Yerka's SF architectures subvert the codes of perspective, multiply points of view and convergence lines, and deconstruct any univocal polarization. The painter complexifies the relationship to space and, by extension, the layers of the ground. Although he often presents cross-sectioned landscapes, his magic universe is opposed to an inertial vision with accumulated strata: It reveals a dynamic agency within the ground layers. In *Double Life*, the dam breaks, showing the connection between the island house and the underwater house; however, the universe of the underwater breaks free from its container to return, through a path, to the Earth's surface (Jacek Yerka, *Boudoir*, this painting can be viewed online at: https://www.wikiart.org/en/jacek-yerka/boudoir). More explicitly, a bucket suspended from the roots in the sky both disturbs and connects the strata of the world of *The Villa* (Jacek Yerka, *Double Life*, this painting can be viewed online at: https://www.wikiart.org/en/jacek-yerka/double-life). Thus, Yerka's topophilia challenges both the strict verticality of architecture and the horizontality of a soil perceived as a smooth, impermeable surface.

The SF architectures claim a thickness and make tangible a conception of the ground, not in terms of verticality, but in revealing its depths: The artist transitions from the verticality of a cross-section to the depths of the drawing. It is not a simple depth revealed to show the truth or the essence of things but rather multiple depths that penetrate the surfaces to propose several convergence lines simultaneously or that thicken both the sky and earth, suggesting material consistency, as seen in *Spring in the Parish* (Jacek Yerka, *Villa*, this painting can be viewed online at: https://www.wikiart.org/en/jacek-yerka/villa). In this painting, the city and its trees form a church. The central axis, among the houses, turns out to be a stained-glass window, rivaling the depth of the horizon line. The city's base sinks into the forest, while the sky is textured, as though by curtains. Such speculative architectural fictions seem to rely on unexpected depths that both breakthrough and thicken the landscapes: They contribute to a reformulation of space, not in terms of geometric extension and discontinuity but in terms of intensiveness and continuity, thanks to the medium of painting. The challenge is then to understand how the fantastical and magical universe of the painter operates this displacement of the conception of space, allowing the intensiveness of the depths to surface (Jacek Yerka, *Spring in the Parish*, this painting can be viewed online at: https://www.wikiart.org/en/jacek-yerka/spring-in-the-parish).

5.3 The Thwarted Perspective

Jacek Yerka's paintings subvert perspective by mobilizing its own rules. Developed in the fifteenth century by architects and artists Alberti and Brunelleschi, who established the principles for geometrically representing three-dimensional space, classical perspective has been crucial in the evolution of architecture and urban planning (Argan & Wittkower, 2004; Cardinali & Perelman, 2018). It relies on three principles: the uniqueness of the viewpoint, which provides, in temporal suspension, a fixed location that organizes the painting; the vanishing point, to which all horizontal lines in-depth converge; and foreshortening, which dictates that objects decrease in size as they recede from the viewer's eye. Yerka fits well into this general framework of classical perspective, which serves more as a paradigm than merely a language or a code (Damisch, 1993). He particularly revisits the pictorial genre of *trompe l'œil* (Calabrese, 2010). For example, in the painting *Paradise in the Yard* (Jacek Yerka, *Paradise in the Yard*, this painting can be viewed online at: https://www.wikiart.org/en/jacek-yerka/paradise-in-the-yard), a frontal view of a courtyard door leads not onto an interior—like the door to number 30—but to an intriguing path at the edge of a wood, prompting the question: Where exactly does the illusion lie? In *Ventimiglia* (Jacek Yerka, Ventimiglia, this painting can be viewed online at: https://www.wikiart.org/en/jacek-yerka/ventimiglia-1992), he paints a rocky coastal landscape, intersected by a car circuit, but here, perspective is disrupted by three dice placed right in the middle of the image, relegating the rest of the scene to the status of a simple game board. Overall, the painter shifts and thwarts expectations in several ways: Vanishing points are no longer unique, the direction of reading relies on ambiguities, and the sizes of figures appear discordant, thus reorganizing the spatial arrangement.

Classical perspective is based on a space governed by the laws of Euclidean geometry. It involves a system of metric relations, anchored by a fixed and stable position assigned to the subject—the point of view—yet represents the perspective of a theoretical mind (Damisch, 1993). The paradox of perspective is that it presents itself at the same time as the point of view of nobody and, thus, anybody, and from nowhere—thus, anywhere. This is what Haraway refers to as the "God trick" (Haraway, 1988), a dominating gaze seeing everything from nowhere. In this "God trick" perspective, space appears as an environment where all elements are mere points within a grid and a system of coordinates. From within, Yerka challenges this linear perspective where uniform convergence lines imply a disembodied eye and a homogeneous space that is itself geometrized. Moving beyond the rationalizing and fixing hegemony of metrics, measurements, and land surveys (Vidalou, 2017), he avoids any omniscient, overhanging, exhaustive, and determined vision to restore space's asperities, ambivalence, and hiding places. The painting *Autumn* is representative in this regard (Jacek Yerka, *Autumns*, this painting can be viewed online at: https://www.wikiart.org/en/jacek-yerka/autumn): In a forest at night, dark trees conceal myriad shelters lit by bonfires in the hollows of branches, trunks, and roots. This mysterious place, where book pages flutter, harbors secret and unseen lives in its nooks and crannies.

Jacek Yerka deviates from the classical perspective by penetrating surfaces and introducing a multiplicity of movements. Through pictorial devices such as openings or passages, the artist assembles or superimposes landscapes, infrastructures, scenes, and angles of view not through a transcendent vision but through limited views with multiple escapes. The composition of *Chisel Landscape* is particularly illustrative in this regard (Jacek Yerka, *Chisel Landscape*, this painting can be viewed online at: https://www.wikiart.org/en/jacek-yerka/chisel-landscape). The central architecture integrates several landscapes with varying scales and lines of convergence. Rivers, valleys, paths, and airflows interact not in a fixed way but dynamically and fluidly, challenging the viewer's gaze. Beyond the abstract and univocal lines of linear perspective, Yerka multiplies the points of view and diversifies the perspectives. In so doing, he develops situated aesthetics (Beauté, 2021), which, based on fundamentally partial perspectives, foster incomplete connections that enable counterintuitive geometries, unexpected openings, and encounters with hesitant views and voices (Frichot & Loo, 2013).

Thus, depth is no longer merely a width seen from the side within a homogeneous and abstract space: It makes matters and landscapes blaze (Merleau-Ponty, 1985). In the work of Jacek Yerka, it is the fantastical architecture, this enigmatic space articulating a heterogeneous plurality of spaces, that ensures the setting in the tension of spatial dimensions: It plays an essential role in arrangement (Deleuze & Guattari, 1980). Thus, the constructions become the occasion and support for a staging that reveals the density and thickness at the heart of the paintings: As in *Chisel Landscape* and like the bioregionalism movement, the infrastructures are embedded in the soil (Sale, 2000). They gather incompossible strata of the terrestrial surface and unveil the worlds beyond—subterranean, submarine, sublunary. Thanks to the SF architectures, the varied depths disrupt the quietness of the surfaces that we know and map, allowing us to see and perceive the other cosmical sides.

5.4 Toward More-Than-Human Lines

The architectural arrangements of Jacek Yerka's paintings dissolve dichotomies—between surface and depth, inside and outside, sky and earth, natural and artificial, animate and inert. By expanding perspectives and movements, the painter nullifies polarities, thus avoiding many reductive dualisms (Plumwood, 2002). The boundaries between up and down blur: The submarine becomes celestial, planets swim in the earth, and clouds float at the bottom of the abyss. Beyond mere continuity between different levels and elements, sky, earth, and sea intermingle. Similarly, the dialectic between outside and inside (Bachelard, 2012) becomes more complex: Not only is infrastructure porous, but the very status of the delimitation is constantly questioned and disrupted, exceeding any form of geometrism. This is the case in *Illegal Manufacturing of Light*, where a wood-burning stove appears to be manufacturing light (Jacek Yerka, *Illegal Manufacturing of Light*, this painting can be viewed online at: https://www.wikiart.org/en/jacek-yerka/illicit-manufacturing-of-light-1993). Behind it, the wall of the room turns into the sky, and a path links the room directly to the horizon outside, without any threshold. The white of the pale sun outside, that of the mist on the horizon, and that of the light from the stove interact to create fantastical connections beyond boundaries.

Walls melt into the sky, cities become volcanoes (Jacek Yerka, *Eruption*, this painting can be viewed online at: https://www.wikiart.org/en/jacek-yerka/eruption), and cars engage in "lizarding" (Jacek Yerka, *Attack at Dawn*, this painting can be viewed online at: https://www.wikiart.org/en/jacek-yerka/attack-at-dawn). Not only are all living beings' chimeras (Coccia, 2020), but the distinctions between the natural and the artifactual, or the animate and the inert, appear to be obsolete. Thus, the painter aligns with new materialisms, which abandon the terminology of matter as an inert substance governed by predictable causal forces. New materialisms propose that materiality is always more than mere matter: It is an excess, a vitality, a difference that makes matters active, creative, and productive (Coole & Frost, 2010).

SF architectures emphasize the synchronic and diachronic thickness of the soil and reveal the temporal texture (Caeymaex et al., 2019) and the historical materiality of the territory (A. Ingold, 2017): Matter embodies its own temporal, historical, memorial becoming. In Jacek Yerka's work, heterogeneous temporal dynamics are at work on the land and its depths. In this context, the painting *Ammonite* depicts a cephalopod fossil that both is in the sea and contains it (Jacek Yerka, *Ammonite*, this painting can be viewed online at: https://www.wikiart.org/en/jacek-yerka/ammonite-1989). Is it the stream that flows into the sea, from upstream to downstream? Or does the ammonite serve as a dam, allowing the stream to flow further downstream? The water's direction is ambiguous. Moreover, the fossil references an extinct species, but in the image, it also resembles a city: Did the inhabitants carve this shape, or did they adapt this structure after the animal's death? The architectures emerge from the soil, emphasizing that architectural "making" is a process of growth (T. Ingold, 2013). The infrastructures, guided by the site, not only arrange the materials without ecological rupture with the milieu but also present themselves as a "making-with," integrating with the thickness of the land and ocean soil, where elements are already agents and partners.

Contrasting with any reduction of nature to mere naturalistic components, SF architectures, in their historical dimension, invite the study of the processes of construction and destruction, of composition and decomposition. They help understand how architectural phenomena weave together multiple layers of temporality to fit within a complex and entangled temporal texture (Naji, 2019). Matters of the ground transform into materials, and these materials return to the earth, the sea, and the sky.

Opposing any fixative and mortifying gaze, this exploration of deterioration calls for particular attention to roughness and asperities, to slowness and connections. It encourages us to think about the relationship between soils and architecture over long timeframes and within open systems (Deleuze & Guattari, 1980; Till, 2009).

Taking the concrete metamorphoses of the soil and the metabolizations of stories in material and collective becomings seriously allows us to move from the unique perspective of technical drawings to more-than-human perspectives (Tsing et al., 2021). Shifting our gaze, Yerka's fantasy world reveals that each entity, whether human or non-human, organic or inorganic, reenacts the world, sometimes reshaping entire landscapes. This is particularly clear in the painting *Port*, where a mollusk shell, surrounded and made by flying or swimming creatures, serves as a port infrastructure and materially connects various worlds—land and sea, living and dead, past and future (Jacek Yerka, *Port*, this painting can be viewed online at: https://www.wikiart.org/en/jacek-yerka/port). Non-humans then become active agents of the biosphere, as they participate in transforming environments in singular and changing configurations: Through sympoiesis (Haraway, 2016), they collectively shape architectural assemblages.

Yerka presents more-than-human agency and autonomy in his work (Prior & Brady, 2017). His paintings consist of trajectories, living and polymorphic lines—conduct lines, lines of life, tracks, departures, and openings (T. Ingold, 2013; Macé, 2019). These lines no longer enclose organisms or serve as contour lines. They create themselves, intersect, evolve together, transform, and take over from each other—as in the painting *Erosion*, where grassy paths intertwine above the city's eroded soil, making the pattern of lines and passages explicit (Jacek Yerka, *Erosion*, this painting can be viewed online at: https://www.wikiart.org/en/jacek-yerka/erosion-2000). Yerka's lines entwine in weavings of interspecific contingencies, with irreversible disruptions and possibilities for recovery (Tsing, 2015). Within such a geography, more-than-human knots thicken the paintings, reminding us that our interactions are contaminating us (Lestel, 2007). Sympoiesis then involves issues of symbiopolitics, i.e., the governance of relations between entangled beings (Helmreich, 2009).

Yerka acknowledges the concrete, embodied, and opaque dimensions of fields and landscapes not by assuming a presupposed living harmony but by producing ground friction. The rubbing surfaces allow us to glimpse and apprehend the living forces (both organic and inorganic) of resistance and resurgence (Tsing, 2011, 2015). However, this work is not about estheticizing the precariousness of the world in a form of apolitical quietism (Lorimer, 2020)—thus, Yerka's SF is not a mere celebration of entanglements in the damaged world of the more-than-human Anthropocene. It does, nevertheless, reveal situations in which entities pursue trajectories beyond human control, embodying the ferality of earthly becomings (Tsing et al., 2021). Fences dissolve into the sky (Jacek Yerka, *Polish Bonsai*, this painting can be viewed online at: https://www.wikiart.org/en/jacek-yerka/polish-bonsai-1996), and houses perch on ice waves that may melt one day (Jacek Yerka, *Winter*, this painting can be viewed online at: https://www.wikiart.org/en/jacek-yerka/winter-2005). Thus, the painter suggests a path toward a conception of architecture that avoids the trap of a hyper-projective society and frees itself from deterministic and controlling paradigms. Life and death lines permeate the landscape, infrastructures become creatures, and the ground's strata engage in friction.

5.5 Geo-Stories of the Depths

Jacek Yerka's paintings propose a shift away from the dominating, totalizing, and centered gaze (Plumwood, 1993), thereby establishing "new regimes of attention" that recognize "here, now, something important is happening" (Despret, 2019, pp. 180–181). This politics of attention, which requires a renewal of our tools and mentalities, is manifest in Yerka's universe through a form of pictorial narrativity, which stages temporalities and becomings. In *Cathedral* (Jacek Yerka, *Cathedral*, this painting can be viewed online at: https://www.wikiart.org/en/jacek-yerka/cathedral), a seashell cathedral with illuminated interior columns seems to float on water. A wooden path leads to it, but the planks have deteriorated. Access is blocked. How did the seemingly present individuals gain access? A closer look and more careful attention reveal a leafy labyrinth, itself inhabited, as evidenced by the small windows, beneath the water. The water becomes earth, a layer is mysteriously added, and the ground gains in narrative thickness.

In a troubled atmosphere, the painter weaves networks of embodied relationships in complex, differentiated, and singular ways. Meanwhile, he responds to the ecological humanities' call to value narrative: To open the imagination and approach the more-than-human modernity in its specificity, we need to tell new stories (Rose & Robin, 2004). This is precisely what these SF landscapes offer, where architectures and structures themselves tend to become decisive characters that significantly disrupt the narrative arrangement of things and environments. This is particularly evident in *Spiritual Hour*, where a large, winged clock takes the role of the painting's narrative subject (Jacek Yerka, *Spiritual Hour*, this painting can be viewed online at: https://www.wikiart.org/en/jacek-yerka/spiritual-hour). By suggesting that architecture and ecology could benefit from layered and varied narratives, Yerka encourages the viewer to follow tangled traces and search for points of view. His new narrative patterns move beyond the dominant gaze's models, plans, and schemas.

Contrary to what the controversial contemporary concept of the Anthropocene suggests, Jacek Yerka does not glorify humanity as a telluric force: "To speak about the Anthropocene is to convince oneself that the Earth is a system that a technoscientific elite must orchestrate, administer and regulate" (Hoquet, 2019, p. 47). Instead, the painter seems to relegate humanity to a marginal role in the Earth's history. In *Theory of Tension*, for example, the gigantic mass of rock suspended in a sunset sky is striking (Jacek Yerka, *Theory of Tension*, this painting can be viewed online at: https://www.wikiart.org/en/jacek-yerka/theory-of-tension). However, while the thin upper layer is covered by a medieval-looking city, this tiny urban surface is not at the core of the pictorial narrative. What truly catches the eye are the rocks stretched out like columns by gravity, the dialectic between the top and bottom of the mass, and the intermingling of the rock strata. Contrasting the systemic and dominant Anthropocene vision of the Earth and the soil, Yerka joins Donna Haraway's "enchanted compostism": "I am a compost-ist [...], we are all compost [...]" (Haraway, 2015, p. 161). Opposing the tightness of modernity and humbly returning to the earth, to the humus, Yerka proposes a geocentric cosmology with porous geo-narratives that radically challenge the perception of the ground as inert and passive. His paintings vividly depict the powerful and dizzying activity of heterogeneous subterranean entanglements, showcasing "the almost-living matter of the earth" (Granjou & Phillips, 2019; Granjou & Salazar, 2019). These geo-stories call for a return to the asperities of earthly matter, emphasizing its concrete, evolving, and living nature.

Grounds and undergrounds are no longer seen as inert receptacles, mere supports for capitalist and neocolonial projects and exploitation

(Vidalou, 2017). Far from imperialism and extractivism, SF architectures decolonize the relationship with the soil and its imaginary. They appear to reject an exogenous domination based on the engineering expertise of modern projects, which often threaten architectural enclosure and patrimonialization. Instead, they seem to promote endogenous, hybrid, and multifocal practices, drawing on indigenous knowledge and the materiality of the site (Naji, 2019). In this sense, Jacek Yerka's magical universe aligns with a decolonial ecology (Ferdinand, 2019), as it rejects territorial ruptures—ecological, landscape, or metabolic—and values networks of relationships. In this framework, the identity of beings is a matter of plasticity, co-composition, openness, and reciprocity among multiple bodies and places. In this respect, the painting *Brontosaurus Civitas* is particularly striking and depicts a gigantic underwater dinosaur (Jacek Yerka, *Brontosaurus Civitas*, this painting can be viewed online at: https://www.wikiart.org/en/jacek-yerka/brontosaurus-civitas). A small part of its back emerges from the water's surface, forming an island where a human city has developed. Closer inspection reveals that the seafloor itself is a labyrinth of buildings, giving the impression of an urban civilization submerged by waters. The painting superimposes periods and stages as heterogeneous temporalities, closely linking very different scales and lives. In so doing, Yerka recognizes the overlapping and kinship of existences without ignoring their independence (Plumwood, 2002): Opposing a colonizing view of soil as a homogeneous, neutral, and thus exploitable surface, the grounds and undergrounds tell stories of terrestrial conditions and more-than-human entanglements.

In these narratives, the map of perspectives and living relationships is not flat but undulating, open, with multiple entries and connectable in all its dimensions, constantly subject to change (Deleuze & Guattari, 1980, p. 20). In contrast to the heroic narrative of the vertical conquest of earth, Yerka's perspectives on the depths reveal stories of geological resistance, "life stories" (Le Guin, 1989): Against the killer story, the Story of the Ascent of Man the Hero, Yerka's life stories tell acentric, nonheroic accounts concerned with the real and the living and attentive to the ordinary yet fantastical course of life. They immerse us in the terrestrial network, create multiple encounters, and reveal unexpected gaps in the evolving cartography of beings. They claim and affirm a power to act and think, intertwine more-than-human lines to disrupt the old cultural sedimentation, differentiate soils and subsoils, and recognize impurities and metamorphoses.

Competing Interests and Acknowledgments This article has no competing interests. I would like to express my gratitude to the artist Jacek Yerka for his works, available on the WikiArt encyclopedia, as well as to the members of the PACTE laboratory for initiating this stimulating reflection on grounds and undergrounds.

References

Argan, G. C., & Wittkower, R. (2004). *Architecture et perspective chez Brunelleschi et Alberti* (J. J. Le Quilleuc & M. Perelman, Trans.). Verdier.

Bachelard, G. (2012). *La poétique de l'espace*. Presses Universitaires de France.

Beauté, J. (2021). Vers des esthétiques situées. In *La beauté d'une ville* (pp. 442–449). Pavillon de l'Arsenal, Wildproject.

Caeymaex, F., Despret, V., & Pieron, J. (Eds.). (2019). *Habiter le trouble avec Donna Haraway*. Dehors.

Calabrese, O. (2010). *L'art du trompe-l'œil*. Citadelles et Mazenod.

Cardinali, P., & Perelman, M. (Eds.). (2018). *Ville et architecture en perspective*. Presses universitaires de Paris Nanterre.

Coccia, E. (2020). *Métamorphoses*. Payot & Rivages.

Coole, D., & Frost, S. (2010). Introducing the new materialisms. In *New materialisms: Ontology, agency, and politics* (pp. 1–43). Duke University Press.

Cowan, J. (Ed.). (1994). *Fantastic art of Jacek Yerka*. Morpheus International.

Damisch, H. (1993). *L'origine de la perspective*. Flammarion.

Deleuze, G., & Guattari, F. (1980). *Capitalisme et schizophrénie. Tome 2, Mille plateaux*. Éditions de Minuit.

Denis, B. (2002). Du fantastique réel au réalisme magique. Textyles. *Revue des Lettres Belges De Langue Française, 21*, 7–9.

Despret, V. (2019). *Habiter en oiseau*. Actes Sud.

Ellison, H., & Yerka, J. (2006). *Mind fields: The art of Jacek Yerka, the fiction of Harlan Ellison* (Gift edition). Morpheus International.

Ferdinand, M. (2019). *Une écologie décoloniale. Penser l'écologie depuis le monde caribéen*. Seuil.

Frichot, H., & Loo, S. (2013). *Deleuze and architecture*. Edinburgh University Press.

Granjou, C., & Phillips, C. (2019). Living and labouring soils: Metagenomic ecology and a new agricultural revolution? *BioSocieties, 14*(3), 393–415.

Granjou, C., & Salazar, J. F. (2019). The stuff of soil: Belowground agency in the making of future climates. *Nature and Culture, 14*(1), 39–60.

Haraway, D. J. (1988). Situated knowledges: The science question in feminism and the privilege of partial perspective. *Feminist Studies, 14*(3), pp. 575–599.

Haraway, D. J. (2015). Anthropocene, capitalocene, plantationocene, chthulucene: Making kin. *Environmental Humanities, 6*(1), 159–165.

Haraway, D. J. (2016). *Staying with the trouble: Making kin in the Chthulucene*. Duke University Press.

Helmreich, S. (2009). *Alien ocean: Anthropological voyages in microbial seas*. University of California Press.

Hoquet, T. (2019). Pour un compostisme enchanté. *Critique, 860861*(1), 44–55.

Ingold, A. (2017). Terres et eaux entre coutume, police et droit au XIXe siècle. Solidarisme écologique ou solidarités matérielles ? *Tracés. Revue de Sciences Humaines, 33*, Article 33.

Ingold, T. (2013). *Making: Anthropology, archaeology, art and architecture*. Routledge.

Le Guin, U. K. (1989). The carrier bag theory of fiction. In *Dancing at the edge of the world: Thoughts on words, women, places* (pp. 165–170). Grove Press.

Lestel, D. (2007). *L'Animalité. Essai sur le statut de l'humain*. L'Herne.

Lorimer, J. (2020). *The probiotic planet: Using life to manage life*. University of Minnesota Press.

Macé, M. (2019). *Nos cabanes*. Verdier.

Merleau-Ponty, M. (1985). *L'œil et l'esprit*. Gallimard.

Naji, S. (2019). *Architectures du bien commun : Pour une éthique de la préservation*. Métis Presses.

Plumwood, V. (1993). *Feminism and the mastery of nature*. Routledge.

Plumwood, V. (2002). Decolonising relationships with nature. *Philosophy Activism Nature, 2*, 7–30.

Prior, J., & Brady, E. (2017). Environmental aesthetics and rewilding. *Environmental Values, 26*(1), 31–51.

Roh, F. (1925). *Nach-expressionismus. Magischer Realismus: Probleme der neuesten europäischen Malerei*. Klinkhardt & Biermann.

Rose, D. B., & Robin, L. (2004). The ecological humanities in action: An invitation. *Australian Humanities Review, 31*(2).

Sale, K. (2000). *Dwellers in the land: The bioregional vision*. University of Georgia Press.

Till, J. (2009). *Architecture depends*. MIT Press.

Tsing, A. L. (2011). *Friction: An ethnography of global connection*. Princeton University Press.

Tsing, A. L. (2015). *The mushroom at the end of the world: On the possibility of life in capitalist ruins*. Princeton University Press.

Tsing, A. L., Deger, J., Saxena, A. K., & Zhou, F. (2021). *Feral atlas: The more-than-human anthropocene*. Stanford University Press. http://feralatlas.org

Vidalou, J. B. (2017). *Être forêts. Habiter des territoires en lutte*. Zones.

CHAPTER 6

In the Ruins of Extractivism: The Unfinished Rehabilitation of the Soil in the Orbiel Valley

Christelle Gramaglia and Sijia Du

List of Abbreviations

ADEME	French Environment and Energy Agency
BRGM	French Bureau for Geology and Mining Research
CGPC	French General Council for Bridges and Roads (Corps of Engineers)
CGM	French General Council for Mines (Corps of Engineers)
COFRAMINES	French Mining Company
GCB	Geochemical background
INRA	French Institute for Agronomy Research (now INRAE—the E being for Environment)

C. Gramaglia (✉)
INRAE, Montepellier, France
e-mail: christelle.gramaglia@inrae.fr

S. Du
University of Beijing, Beijing, China

Institute of Political Sciences, Paris, France

© The Author(s), under exclusive license to Springer Nature Switzerland AG 2025

O. Labussière et al. (eds.), *Back to the Ground*,
https://doi.org/10.1007/978-3-031-88888-5_6

MOS	Salsigne Gold Mines Ltd.
RMQS	French Soil Quality Measurement Network
SMPCS	Salsigne Mines and Chemical Company Ltd.
SEPS	Salsigne Exploit Pyrometallurgy Company Ltd

6.1 Introduction

The Mines d'or de Salsigne (MOS) in the Orbiel Valley, north of Carcassonne in southwest France, were closed in 2004. This closure not only sparked ire among the miners and metallurgists made redundant but also raised concerns among the local population about pollution at the site. Initial studies indicated the extent of the rehabilitation work required to deal with the vast volumes of waste generated by gold and arsenic extraction over more than a century. Several companies operated the site, including Société des Mines et Produits Chimiques de Salsigne (SMPCS), COFRAMINES (a company belonging to the French geological and mining research bureau, BRGM), MOS (owned by Eltin Ltd. and Sons of Gwalia Ltd.), Société Exploit Pyrométallurgie de Salsigne (SEPS), and Herbinger-SNC Lastours Ltd., each using different technical processes. Initially, there were mine shafts, followed by an opencast mine in the Orbiel Valley with facilities for industrial hydrometallurgy and pyrometallurgy. The resulting waste—waste rock, slag, and fine residue—was stored, not always in a stable or controlled way, or simply left abandoned on site. As early as 1995, following the bankruptcy of the operator SEPS, a report by the Conseil Général des Mines (CGM) and the Conseil Général des Ponts et Chaussées (CGPC) for the ministers of the environment and industry noted that creating a "complete inventory" was impossible due to the sheer volume of waste. The long history of mining and metallurgy, as well as the succession of operators, makes it even more difficult to assess the situation accurately. This, in turn, causes concerns to arise, as in 2018, when flooding drew attention back to the issue of pollution because contaminated sludge had to be removed from the residential areas.

Despite an attempt from the 1990s on to restore the site at a cost to the State of over €45 million, the problem remains unresolved. Questions persist about the effectiveness of the work undertaken by the Agence de l'environnement et de la maîtrise de l'énergie (ADEME), as well as the ongoing monitoring by BRGM. Indeed, the chain reactions caused by what Nigel Clark calls "de-stratification" (2014; Clark & Richards, 2022) do not cease once mining activities are suspended, especially when the process of "turning over the Earth's crust" has been harsh, releasing toxic materials that had formed over geological time in other ecological contexts (Ureta & Flores, 2018). Moreover, slag, waste rock, and fine tailings are not as inert or controllable as we are often led to believe. They have a tendency to spill out of their containment areas, disperse, and spread—in other words, to pop up unexpectedly, causing both environmental and social disturbances (Boudia et al., 2018). The volume of this waste far exceeds the quantity of valuable ore extracted from it. For example, it is estimated that approximately 100 tons of gold and 400,000 tons of arsenic were mined in the Orbiel Valley. This process involved excavating over 20 million tons of material, which was stirred up, brought to the surface, and processed to separate elements previously bound together in a geological gangue that kept them relatively stable. Some arsenic was sold, but much remained at the site as waste in various chemical forms, mixed with other toxic elements, including metals and organic pollutants from the incineration of industrial waste in the SEPS furnaces during the final months of its operation.

Considering the extent of the problem, what can we expect from rehabilitation efforts? What can the Orbiel Valley teach us about techniques and practices in this field? To what extent can we "clean up" after such a destructive process of geological de-stratification? After outlining the value of human and social science approaches to post-mining problems, we review the chronology and nature of the work undertaken in the Orbiel Valley. We then discuss recent disputes to analyze the epistemic cultures (Knorr-Cetina, 1999) and sociotechnical imaginaries (Jasanoff & Kim, 2019) of those involved in rehabilitation. How did they see the outcomes of their work and the future? What are their objectives and resources? How do they address the recalcitrance of the materials they intend to neutralize? We focus on a dispute over the geochemical background that allowed BRGM engineers to make the arsenic levels in the soil appear normal, against the advice of their colleagues at Toulouse University. Why did their calculations differ? What are the consequences in terms of time projection, reversibility, and responsibility? The aim of this article is to spark a discussion on mining liability at a time when the reopening of mines is being considered in France and the European Union.

6.2 Toward a Sociology of Post-mining Management

Historically, sociologists have overlooked the "underground," except through surveys of miners' work that dealt with it indirectly, particularly following the closure of major mines in countries in the North and their development in the South during the 1980s and 1990s. It was not until the recent mining revival, spurred by strong global demand for raw materials, that studies were revived as a result of opposition to new extractive projects (Le Berre & Chailleux, 2021) and disputes over the management of mining liabilities (Bonincontro et al., 2023; Hecht, 2023; Mésini, 2018). This rediscovery of the underground by sociologists has been shaped by *science and technology studies*, or *STS* (Kinchyet al., 2018), which portray soil and the underground not just as inert and passive substrates for human activities but as complex assemblages of material and biological entities with specific relational forces (Gramaglia, 2020). They have been objects of a social practice that has promoted their requalification as resources for exploitation. These new studies have emphasized waste, its producers, the harm it causes, and the sociotechnical reconfigurations needed to tackle its proliferation, raising questions about its potential deep disposal.

This shift has led to the most insightful reflections on the effects of geological de-stratification mentioned in the introduction (Clark, 2014; Clark & Richards, 2022). This process can be understood literally as an excavation operation—a "turning over of the Earth's crust," "disruption of rock tissue," or "destabilization of the Earth system." It also highlights the role of the soil and the underground in the advent of the Anthropocene. The literature on this subject extends beyond mining to include oil in discussions about the causes of climate change. However, mining is particularly suited for an analysis of de-stratification due to its brutal transformation of the landscape and enduring impact on ecosystems and their human inhabitants. For example, geographer Sigrun Kabish (2004) describes the daily experience of the ravaged environment around the former brown-coal pits of East Germany, which underwent one of the biggest remediation programs in reunified Germany. By "remediation," a technical term synonymous with rehabilitation, she meant the extensive, costly efforts required to clean and secure sites after a mine's closure. Sociological studies of these efforts often emphasize the ignorance and uncertainty surrounding the decisions made. They also note the urgency of actions taken, with actors not waiting to find out anything that could prevent the process, despite knowing that "many things can go wrong" (Gross & Bleicher, 2013, p. 200).

This argument resonates with research on the indigenous land of the Yellowknife Denes in northwestern Canada, where a gold mine named Giant Mine operated from 1948 to 2004. Following the bankruptcy of the operator, authorities were slow to take action to restore what had become one of the most polluted sites in North America. Current efforts involve freezing 237,000 tons of arsenic trioxide dust underground, to be monitored for at least 100 years. Other storage sites for less hazardous waste are planned above ground. Geographer Caitlynn Beckett notes that in Canada and elsewhere, the approach to rehabilitation is excessively technical, which has the effect of staving off demands from affected communities for apologies or compensation (2021). Such measures never fully restore integrity or the previous safe environmental state but aim to transfer mining waste to special landfill sites or stabilize and contain it on site. At best, they allow certain biogeochemical functions and services to recover, but they do not facilitate learning about the extent of the degradation or its implications for the residents. Local communities are excluded from the decision-making process. Thus, Beckett finds it neither possible nor desirable to speak of restoration or remediation—while some authors prefer the second term for its medical connotations, suggesting a multidimensional healing process (Cerceau & Laurent, 2023). Instead, Beckett consider that the situation calls for an obligation of perpetual monitoring, with recognition of the inevitable leaks and overflows.

This view, however, is not shared by most scientists, particularly engineers in the rehabilitation sector, whether publicly or privately contracted. They are inclined to technological optimism, to borrow an expression by Anthony Giddens (1990). To understand the origins of this attitude, we explore the notions of "epistemic cultures" and "sociotechnical imaginaries." These frameworks help elucidate how those involved in rehabilitation—scientists, engineers, and technical staff of companies in the sector more generally—think and operate.

According to Karin Knorr-Cetina:

> The notion of epistemic culture is intended to capture the internalized processes that count in the production of knowledge. It refers to that set of practices, arrangements and mechanisms which, bound together by necessity, affinity and historical coincidence in a given field of professional

> expertise, ensure that we know the way we know. Epistemic cultures are the cultures of knowledge creation and validation. (2007, p. 363)

These cultures influence many choices that guide research and risk assessment—identifying subjects, devising methods, and interpreting results—and thereby shape the evidence by shedding light on certain aspects of the issues under investigation while obscuring others.

Combining epistemic cultures with the more recent notion of sociotechnical imaginaries (Jasanoff & Kim, 2019) focuses on future forecasts for managing waste whose harmfulness and lifespan far exceed the usual spatial and temporal scales. Assumption, projection, and conjecture are needed to justify the selection of technical options in the face of uncertainty and limitation. The National Methodology for the Management of Polluted Sites and Soils, introduced in France in 1993 and revised in 2007, advises planning rehabilitation work based on use and risk calculations, within so-called "economically acceptable" limits. The goal is not to "clean up" pollution, if that were even possible, but to limit transfer and exposure at a reasonable cost. Therefore, pollution must be contained, whether by drying up the source and deactivating dissemination routes through physical, chemical, or biological stabilization techniques or by excavating and confining it in controlled storage. However, neither approach is foolproof. Each assumes a timeframe within which the reliability of the processes and equipment is considered adequate and establishes a monitoring period to ensure this. In all cases, technical decisions are based on assessments and modeling that select certain variables deemed decisive in anticipating and shaping "acceptable" future conditions, based on known criteria and factors. Anything unpredictable is not considered, leaving successors to manage any issues that arise.

As "prospective visions" (Sismondo, 2020), sociotechnical imaginaries are not homogeneous; like epistemic cultures, they vary based on the training and agendas of those involved. Some may even compete with each other and give rise to disputes. Once an imaginary has succeeded in establishing itself, it reinforces a particular social order, in which technology—specific techniques in particular—plays a crucial role. It is important to note that, if not made explicit, the highlighting or obscuring effect of each epistemic culture and sociotechnical imaginary tends to enact one type of future when other, possibly more favorable, options could have arisen—especially regarding social and ecological injustice

(Granjouet al., 2017). Our study aims to reopen the question of the alternative futures that could emerge if we employed other methods and instruments to manage mining liabilities. This is not to suggest that we should have acted differently, but to explain that none of our options is ever neutral or solely technical. Each option carries social and political dimensions to be made explicit.

6.3 Methodology

The first step in our investigation was to consult archives, specifically those made available by the *Gratte-Papier* advocacy group. This group had collected all the official reports and documents concerning the management of the mine and pollution in the Orbiel Valley, including technical reports, prefectural rulings, publications by associations, and press articles. We focused on the information regarding the rehabilitation and monitoring of the site in its post-mining era. We also looked at government websites and conducted 15 lengthy, semi-structured interviews with various pollution experts: engineers from government agencies (ADEME and BRGM), scientists from the universities of Montpellier and Toulouse, health professionals, mayors, and representatives of local associations. The recordings were transcribed, coded, and analyzed. This empirical data was supplemented by observations from several public meetings in the Orbiel Valley and "technical" visits to rehabilitated sites.

It should be noted that, despite repeated requests, the public offices for the environment at the regional level either did not respond or declined to be interviewed, compelling us to rely on their official statements to understand their position. This lack of responsiveness is indicative of the atmosphere of conflict, secrecy, and mistrust that characterizes the Orbiel Valley, making it a particularly challenging site for study.

6.4 Rehabilitation Objectives and Conditions

A Division of Labor that Conceals the Absence of an Overall Plan

The mining site we studied is situated around the Orbiel Valley—a tributary of the Aude River that flows into the sea at Narbonne. The site includes Salsigne and other communes in the Montagne Noire region, such as Lastours, Villardonnel, and Villanière, covering a total of seven concessions spanning approximately 3980 hectares. The area includes mine shafts, an opencast mine, waste dumps, brownfield sites where ore was crushed and extracted, and today, storage sites for waste rock, slag, and fine tailings, as well as ponds for treating or evaporating contaminated runoff water. Despite this extensive history, no one visiting the valley today would guess its past, as it is now dedicated to wine production and tourism. Apart from a few signs prohibiting access, fences, a mine trestle, and a memorial to the miners, the rehabilitation work carried out in the 1990s and 2000s has largely erased the most visible traces of the past, and now the hills are covered in vegetation.

Yet, mining activity in the Orbiel Valley dates back thousands of years and has shaped the landscape. As early as the second century BC, the Romans were extracting copper. However, it was not until 1892 that gold was discovered. Mining expanded throughout the twentieth century, at a faster or slower pace depending on the price of precious metals and costs and techniques of extraction. The opening of a large opencast mine in the 1980s marked both the peak and the end of this process.[1] In 1979, SMPCS, the operator of the historic gold mines owned by Silver Eureka Ltd., faced difficulties. To preserve jobs, the government requested BRGM to become the main (public) shareholder via its dedicated operating subsidiary, COFRAMINES. Ten years later, the company still required restructuring. SMPCS's mining activities were sold to MOS, while SEPS took over the smelter to incinerate waste for recycling. Meanwhile, Herbinger-SNC Lastours Ltd. was tasked with reprocessing waste rock and slag, which might still contain gold, by cyanidation. Prior to the site's definitive closure in 2004, various companies intervened to salvage some of the business while carrying out the initial safety and rehabilitation work. However, these efforts were abandoned after a few years, which complicated the task that the State subsequently had to undertake (Cour des Comptes, 2003).

While MOS gathered some of the SMPCS mining waste at the Artus tailing and in the Montredon basin, the public body ADEME dismantled the Combe du Saut metallurgical plants. This division of labor was established to manage the waste with the prospect of reusing some of the waste rock, slag, and fine residue while regulations were still developing. Given the early failures of the operators, it fell to BRGM, under

[1] For a fuller history of the mining site and its conflicts, see Hervé Pujol, *Tristes mines, impacts environnementaux et sanitaires de l'industrie extractive* (LEH Editions, 2014) and Alexis Jaoul's current thesis.

its Mining Prevention and Safety Department, to look after the other, smaller abandoned sites. The same public body was then commissioned to monitor the storage sites indefinitely, as some waste, including the Orbiel riverbanks and the open pit, has still not been treated. Thus, the rehabilitation process appears surprisingly disjointed and lacking in coordination, as pointed out by a government agency engineer involved in post-mining management, who lamented the lack of planning.

> There was no overall management plan. That's the problem! [...] There was no overall management plan for the area. There have been management plans for certain areas. What we did—and this didn't exist at the time; it wasn't called a management plan—was a study of the sector we were responsible for. After that, we weren't in charge of the river, or the soil downstream, or any of that. (Yvan, engineer)

As he explains, his agency dealt with a sector that was regulated concerning environmentally classified installations and strictly adhered to that sector. In addition, it outsourced several studies and technical operations to specialist consultancies and other companies, thereby increasing the number of professionals involved. This compartmentalization of rehabilitation processes and the targeted environments, as well as the involvement of multiple agencies and departments, has led to a lack of accountability. For example, engineers often answer to queries by saying the problem falls outside their remit or area of expertise, such as Alice, who commented on sediment in the Orbiel: *It's not our responsibility.* (Alice, engineer).

In the past, ADEME made technical proposals for the €27 million project to rehabilitate the Combe du Saut, and the prefect (local authority) and regional government departments made the final decisions. Contractors were hired; factories and furnaces were demolished. According to the operation's supervisor, the soil was "stripped" over an area of around 30 hectares. Materials contaminated with over 3 g/kg of arsenic were collected in an old valley, already filled with waste, on a shale substrate. The whole area was then covered with a bitumen membrane. Phyto-stabilization experiments were carried out on the excavated plots in an attempt to limit erosion and the transfer of residual arsenic, which remained too high for most uses. Some plots were covered with iron filings and then planted. This saw limited success, and there was no decision to expand the operation. Sparse vegetation has since grown on the otherwise undeveloped site. The interview excerpt below refers to the decision taken at the time not to add healthy soil but rather to try to stabilize the plant cover while minimizing exposure.

> In fact, we ended up with soil containing around 2.5 g of arsenic per kg of soil, which, incidentally, is still high with regard to health issues. But we're in a sector where there was no activity... If we'd wanted to cover it up, we'd have had to bring in very large quantities of soil... We don't know where we'd have got it from... Phyto-stabilization is a technique that we're continuing to try to use as a means not of solving the whole problem but of reducing the risk. (Yvan, engineer)

Thus, the goal was less about rehabilitation and more about "reducing the risk" by containing most of the waste and stabilizing the remaining contaminants in the "stripped soil." The idea was also to improve the way the site is perceived. This moderation of ambitions can be found in many technical reports and documents. Indeed, as early as 1998, a report to the ministers of the environment and industry pointed out that the cost of decontamination could be enormous—even unaffordable.

> The objective of completely decontaminating the site is unrealistic because the cost of the operation would be extremely high. It's not even possible to assess the cost of this solution until extensive studies have been carried out on soil pollution, which will only be possible once the masses of waste and fill on the surface have been removed. Therefore, we propose that the objective be to rehabilitate the site as defined at level 2, namely, to take all necessary steps to ensure that the waste on site no longer causes any significant pollution outside the site and to restore the image of the site. (CGPC, CGM, 1998, p. 5)

Some of the waste was so "arsenic-rich" that it could not be sent to a class 1 industrial landfill, of which there are a dozen in France, including two relatively close to the Orbiel Valley (at Graulhet in the Tarn and Bellegarde in the Gard). It was decided to send it to Stocamine in the Haut-Rhin, where a class 0 landfill had just been created in an old salt mine for waste too hazardous to recycle. The engineer quoted above confirmed this.

> Afterward, the factory was very polluted, because we had products... it's "pure," you know; pure arsenic! You could call it that. And so we

> don't know what to do because there's no way of dealing with it... The waste from Salsigne was so arsenic-rich we couldn't put it in a landfill for hazardous waste in France. (Yvan, engineer)

However, the Stocamine solution was terminated due to a fire and financial problems, meaning waste has been left on site in the Orbiel Valley for several years. This includes the lime arsenates generated by the Combe du Saut water treatment plant, which were intended to be temporary but remain active due to the high concentrations of metals in the water. An additional watertight cell was built to receive the sludge, but it quickly filled up. In subsequent years, the sludge was bagged and stored in the open air on a neighboring plot of land until 2020, when the prefect, under pressure from advocacy groups, agreed to remove it.

MOS, meanwhile, contained most of the other waste in existing dumps—the least toxic on the Artus tailing and the most toxic, with high concentrations of cyanide arsenic, in the Montredon basin. The company received €4.2 million of public money for this work. It was subject to the mining code, which lacks environmental provisions and is even less restrictive than the regulations for classified facilities. Consequently, to save costs, MOS negotiated a reduced thickness of the initial clay layer and a geotextile cover for its Montredon storage facility, which it claimed would remain watertight for 150 years (it lasted about 10). It should be noted that MOS was a mining company with expertise in extraction rather than rehabilitation and was focused on cutting costs. As one expert observed:

> The Australians were miners. They suggested [their solution] to the authorities, who said OK. [...] It's cheaper. And then they did it. In the lab, it works well. But it doesn't work in practice. [...] They weren't very good at MOS. I don't think rehabilitation is their business at all. And they did as they pleased. There should have been more oversight. There would have been fewer problems. (Yvan, engineer)

Numerous defects and malfunctions were later identified in the MOS storage facilities. Not only did the permeability issue at the Montredon basin lead to water pollution, but the very stability of the building was compromised. Designed to hold 2.1 million tons of waste, the basin was loaded with almost 2.7 million tons, with no penalties or corrections imposed by the authorities. It fell to the new site manager, BRGM, to undertake remedial works, addressing problems as they were identified, often through reports from local associations or residents who noticed anomalies, such as acid mine drainage in ditches or on roads.

Cost-cutting Leads to Less Sustainable Solutions

Our observations show that the various solutions were based on a simplified objective: not to clean up but to minimize the risk and its perception (see Table 6.1 below). Government departments initially focused on recycling and recovery, which had the advantage of retaining and even creating jobs. This approach, reminiscent of the "invisible hand" notion—whereby the market and competition are expected to resolve issues—proved overly optimistic. Nonetheless, the re-mining of the waste to extract residual gold provided an opportunity to establish the first controlled storage facilities. Businesses were attracted by the prospect, particularly the government grants. They consolidated the waste to contain the toxic materials and prevent them from spreading—at a reduced cost. The entire approach was predicated on cost-cutting. For example, the geotextile that MOS placed over its storage facilities had a very short lifespan. Why did the government not intervene to mandate the use of bitumen, as recommended by ADEME? What prevented them from stepping in when the storage capacity was exceeded? The entire operation compromised efficiency; yet, expenditure on site maintenance only increased since the 1990s as new problems were detected. This is disproportionately high compared with the wealth produced by the gold mines in the Orbiel Valley, most of which was captured by private operators, while remediation costs are borne mostly by the public.

ADEME also opted for excavation and containment under bitumen but incorporated phyto-stabilization. Iron filings were spread and followed by seeding to trap arsenic, storing it in plants rather than degrading it, although the method had not yet been proved effective. The state engineers wanted to conduct this research and primarily sought funding for their work—here through a LIFE program from the European Commission. Their aim was to test new methods on a large-scale pilot site, which could be replicated elsewhere. Several plots were equipped, but despite promising outcomes, the experiments were not continued. Some of the equipment can still be seen abandoned in the field, lying on a roadside embankment, overgrown with sparse vegetation. All plots are subject to restrictions on use. This approach was defined with

Table 6.1 Presentation of rehabilitation methods in line with an objective

Rehabilitation methods	*1*	*2*	*3*
Technical	Re-mining/ recycling	Excavation/ containment	(Phyto-) stabilization
Operators	Companies	Companies and/or the State	State and scientists
Objective	Creating value from waste	Secure storage on site or in a special landfill to isolate waste	Fixing the pollution to contain it by using chemical or biological interventions

the aim of creating a demonstration site—not so much implementing a solution.

In addition, the results of monitoring the water treatment plant have been unsatisfactory. Monitoring was intended to be temporary, as the flow of arsenic into the Orbiel carried by runoff water was expected to be reduced by between 75 and 90% (i.e., from 2,500 tons/year to 500 kg/year). However, the flow is still estimated to be 8 tons/year, despite the work undertaken over the years (Cour des Comptes, 2003), and the planned extensions of phyto-stabilization tests have been abandoned. The technical containment and stabilization systems have proved largely insufficient, and the waste has continued to spread. Local associations have described this as a "pharaonic project of questionable effectiveness."

Moreover, rehabilitation was planned based on stable climatic parameters, without worrying about the irregular rainfall, the Orbiel River that can overflow and spread arsenic, or the fluctuating groundwater levels in the Montagne Noire, which can lead to increased acid mine drainage. At no point were any provisions made to account for the effects of exceptional events, such as hundred-year floods. Such measures were deemed too expensive. Once again, cost-cutting is the guiding principle, and uncertainties are minimized. The management of potential hazards is postponed, delegated to future "repair" actions, as the following testimony illustrates:

> Admittedly, it's difficult to find solutions that can cope with rainfall over 100 years or with exceptional events. The Combe du Saut was designed for a 10-year event… This means that if there's a 100-year event, the

> water flows everywhere: The gutters are submerged. Repairs will have to be made. The networks will be inadequate; that's for sure. You have to accept that.... (Yvan, engineer)

The erratic flow of water, dramatically exacerbated by climate change, is barely acknowledged in calculations until it causes an overflow that must be addressed. The relational forces of waste in the presence of new elements, such as water or oxygen, are largely overlooked, despite oxidation transforming it to the point of increasing its toxic power. The focus remains on inexpensive methods to quickly resolve problems, even if it means relying on temporary solutions that will soon require corrective action. There is confidence that future science and technology will bring innovative solutions, allowing other interventions and expenditures to be deferred. This approach, tied to specific epistemic cultures and sociotechnical imaginaries fueled by technological optimism, implies that substantial monitoring and maintenance costs will be incurred indefinitely, involving regular patching-up and sealing work, as well as recurring disputes (at the current cost of 500,000€ per year for the French government, allocated to BRGM and its subcontractors).[2]

6.5 The Geochemical Background Dispute

From Prospecting to the Classification of Geological Anomalies

The geology of the Montagne Noire, formed from the friction between the Massif Central and the Pyrenees, has abundant resources of minerals. Some interviewees described it as a favorable "geochemical anomaly," yet there is concern that excavating elements less profitable than gold, some of which are toxic, has led to numerous environmental and health problems. De-stratification has produced massive amounts of waste that, within just a few decades, have disrupted geochemical balances that took thousands of years to form.

The concentration of elements, particularly metals, is not uniform in the soil and underground but varies over short distances. For mineral prospecting, identifying "anomalies" promises profits when valuable ore

[2] By way of comparison: the Orbiel Valley mines produced 120 tons of gold and have cost over €45 million to rehabilitate, while Giant Mine in Canada produced almost 200 tons of gold and will require work worth between USD 1 and 4 billion. Both sites will require perpetual monitoring and care.

is discovered. In rehabilitation, the question is whether the "anomalies" are natural or man-made. The impact of mining must be ascertained to properly calibrate the work, which does not aim to completely remove all contaminants. Arsenic, for example, is naturally present in the Orbiel Valley. Even before it was mined, it was already in the soil at higher levels than typically found elsewhere, outside of the Montagne Noire and other specific rock formations. Therefore, the goal is to eliminate the surplus caused by mining and industry and bring it to a level comparable to the geochemical background (GCB), which serves as a reference point from which to assess an alteration and plan its repair (Ureta, 2018). Scientists define this target state as:

> A theoretical concentration range of a substance in an environmental sample (or matrix), considering its spatial and temporal specificity. (Gałuszka & Migaszewski, 2011, p. 9)

The approach involves identifying anomalies relative to the past or a geographically close geochemical reference point, where human disturbance is minimal. Various sampling and calculation methods can be used to differentiate a normal from a problematic state, as there are no quality standards for soil in France or Europe. It is rehabilitated not according to a threshold but in line with conditions that would exist in the absence of extractive or industrial activity. This leaves room for interpretation, and, consequently, disputes arise when experts or residents disagree, as explained below.

> There are several methodologies for determining the geochemical background, and that's a bit of a problem; people don't necessarily agree on a common methodology, and as a result, the values we find can change depending on the method. (Isabelle, scientist)

In the Orbiel Valley, the initial objective of the rehabilitation, for example at Combe du Saut, was to excavate and contain all waste with arsenic concentrations in excess of 3 g/kg, which is still very high, but the area is mostly uninhabited so the risk may be considered less. Since the residue could not be "cleaned," it had to be stabilized to limit its spread. The question of the GCB arose later when BRGM had to implement additional measures to reduce the transfer of arsenic and other contaminants into the environment and from the environment to living beings. As a baseline, its engineers used an inventory carried out between 1975 and 1992 to locate seams to mine in response to a crisis in the national supply of raw materials (not for a soil rehabilitation operation). However, after statistical calculations, it was possible to establish a range for the Montagne Noire geological so-called "natural" anomaly for arsenic of between 45 and 339 mg kg^{-1} (BRGM, 2021). The method was designed to accurately represent the area's soil by dividing it into three distinct geological units. Based on numerous samples, including those taken near mining sites, it also accounted for the specific mineralization processes occurring there. Yet, while the volume of data was intended to guarantee the objectivity of the diagnosis, one result of the approach was to "normalize" the high concentrations: Certain peaks were reintegrated into the natural order. BRGM is careful not to overestimate the risks of anthropogenic origin because the soil in the Orbiel Valley contains arsenic just as elsewhere in France. As one engineer said,

> It's completely natural. That is to say, in France, the average level of arsenic is naturally between 20 and 250. (Alice, engineer)

The advantage of this approach is that it makes the task easier—or at least possible and affordable—for those managing the post-mining process. It allows them to focus their efforts on a reduced number of sites requiring special attention while overlooking the others. The problem arises when the risk is underestimated, which was evident following the floods of 2018 when large quantities of contaminated material were washed away. Consequently, geochemists at the University of Toulouse found it necessary to specify the local GCB. They took samples from the wooded part of the Montagne Noire outside the mining area to calculate an average level from non-impacted sites. This explains the difference between their results and those of their BRGM colleagues at the Orléans research center.[3] They determined a precise value: 44 ± 12 mg/kg for arsenic (Delplace et al., 2021), questioning the adequacy of a range they viewed as insufficiently cautious for local populations.

[3] It was undertaken not by BRGM's "Unité territoriale Sud," which is responsible for post-mining management in Gardanne's Orbiel Valley, but rather by a group of researchers affiliated with the "Direction nationale des ressources minérales" in Orléans.

Different Methods Lead to Different Results

Shortly after publishing their article in *Chemosphere*, an international scientific journal, these geochemists were surprised to be contacted by the publisher, who informed them that the BRGM team, authors of the first report on the GCB in the Orbiel Valley, had requested a right of reply. *Chemosphere* published the BRGM response (Melleton et al., 2022), which criticized both the sampling method and its reference to the Réseau de Mesure de la Qualité des Sols (RMQS) of the Institut National de la Recherche Agronomique (INRA, now INRAE). This is based on standards from the French Ministry for Ecological and Solidarity Transition for polluted sites and soil. The BRGM deemed the methodology irrelevant for not directly addressing exposure and health risks, and the statistical calculations were also criticized for being insufficiently descriptive, meaning the spatial distribution of arsenic concentrations remained unclear. The authors emphasized the natural presence of arsenic at high levels and the need to acknowledge the degree of variation in concentrations at the same site.

The same team referred to a second report based on new samples and soil analyses with slightly different values from a study carried out to clarify the GCB in the Orbiel Valley (BRGM, 2021). This time, the level was between 32 and 520 mg/kg. Being both lower and higher than the scientists' findings, this range again normalized the geological "anomaly" of the Montagne Noire. Natural concentrations are estimated to be low, with possible excesses and occasional peaks not linked to mining, giving an average of 123 mg/kg—almost three times higher than the value used by geochemists in Toulouse.

The opposition between the two teams reflects their respective agendas and responsibilities. The former seeks to establish an unambiguous GCB and, thus, needs a pollution-free zone to establish a baseline. By contrast, with a management approach, the latter aims to represent the locally observed variation, even if it introduces uncertainty over what is natural and what is human-made. A researcher from the Toulouse team highlighted the difference. In her view, the geochemical background must necessarily refer to land that has not been altered by extractive processes if the impact of mining is to be truly assessed.

> We're always trying to determine what's natural, so that means staying far away from mining activities. Natural geochemical background really means

> staying away from all man-made activities. Whether it's farming, building a road, building houses. That's really what it's all about [...]. They [BRGM] got the job done—and we're not criticizing that. Their work was done for mining prospecting purposes. So, in fact, they took samples that they considered to be geochemical background with the aim of being able to drill and find gold. They thought about their sampling strategy, instead of moving away from the mining district, agricultural areas, and so on. They went right to the heart of where they thought there was the most! That's where we find the bias, because, of course, they have values of around 100, 150 mg per kg. It's not a huge amount compared with us, but it's still three times higher.... (Stéphanie, geochemist)

The Toulouse geochemists, in turn, requested their right of reply (Delplace et al., 2022), and methodically addressed each of the criticisms leveled at them by BRGM. They began by contextualizing the disagreement, noting that their GCB was almost equivalent to the lower value of BRGM's range and that the lowest value should be used to protect the public. They explained that their calculations were based on the RMQS and references for polluted sites and soil. Lastly, they called for more direct exchanges and, above all, for the latest BRGM study to be reread and peer-reviewed, which is not obligatory for government technical reports. The excerpt below highlights the similarities between the two teams and the social and political implications of risk assessments, urging all parties to assume their responsibilities.

> The results of our two studies seem to be consistent. Indeed, our value for the pedogeochemical background determined in the Orbiel Valley for arsenic (As), at about 44 mg/kg, remains within the range of the low value reported in this engineering report. In risk assessment, choosing the lowest value may be preferable to better protect populations and ensure that the risk is thoroughly investigated and considered. (Delplace et al., 2022)

It was notable that the exchange of views took place in a scientific journal, given that the teams were respected professionals who knew each other. They had crossed paths in 2022 while presenting their research at a study day on arsenic at former mining sites. The dispute played out in writing and indirectly through public arenas, via associations, advocacy groups, and regional government departments, all using the data available to support their arguments. However, a kind of hostile ignorance poses challenges in understanding the stakes involved. Consequently, many of those we approached refused to answer our questions. BRGM's official position on the GCB had to be inferred from the letters and comments previously mentioned. The only reaction we received, off the record, was that the article had been perceived as "incriminating" against the State agency, which the scientists refuted.

One explanation is that the Toulouse geochemists sought to establish the GCB by taking samples at the surface and in-depth in areas strictly unaffected by mining, industry, or agriculture, while BRGM may have sampled at the surface only but across many points near former mining sites. In the first case, it is acknowledged that the choice of sampling method has immediate implications for the local population.

Quality land used differently from the rest of the valley is targeted. In the second case, the methodology is more conservative and guided by a standard for already polluted soil. For representativeness, samples are taken closer to the mining sites, risking an overestimation of the original arsenic levels and thereby concealing the human impact. They rely on existing data supplemented with measurements from a field fluorescence XRF gun—reliable, yet less precise than laboratory tests and ICP-MS analyses. Each method has its own specific features, as set out in the Table above (Table 6.2). They reflect different approaches to data production and understanding the social implications of expert knowledge. They also have different effects on the states of the world they describe and ultimately help to enact, with unequal abilities to accept their responsibilities in this regard.

6.6 Conflicting Epistemic Cultures and Sociotechnical Imaginaries

The rehabilitation of the former mining site in the Orbiel Valley allows us to examine at least two contrasting, though occasionally similar, epistemic cultures underlying research methods. In the first part, we saw different approaches to waste containment and varying results. For both MOS and ADEME, cost considerations govern technical choices; MOS looks to the market, receiving public subsidies and keeping costs down, while ADEME seeks a solution that can be transposed to other polluted sites. The relative effectiveness of their input is clear: MOS has gone into insolvency, and ADEME was replaced in its monitoring operations by BRGM. At no time in any of the documents or interviews we gathered did anyone express uncertainty about or suggest limitations to their methods, nor

Table 6.2 Summary table of the different methods of obtaining a GCB

Actors	*Geochemists*	*State experts*
GCB	Several ways to calculate it	Single approach
Reference	Scientific and technical methods in connection with quality standards for soil	Standard method for polluted soil
Sampling	In forests never logged and away from agricultural areas + upstream of rivers (extrapolation to sediment), in-depth	In contrasting soil zones representative but also near mining and industrial sites + along riverbanks at the surface
Analyses	Laboratory acid tests on samples and ICP-MS	Mining inventory data and XRF on-site measurement
Result	Choose a low, cautious single value—with information on uncertainty range	Choose a value range—without information on the uncertainty range
Special feature	Precautionary and reflexive approach in relation with the social and political implications of calculations	Approach eliminating outliers—circumscribes a reduced scope of intervention for the managers

have authorities invited them to reflect on possible failures. They all maintain a technological optimism that believes later corrections or solutions can be implemented within a narrow concept of approach and timeframe. For example, a geotextile can be replaced without major consequences, apart from potentially infinite monitoring. This sociotechnical imaginary starkly contrasts with that of the Orbiel Valley residents, whose attachment to the land drives their desire for sustainable responses that last generations and demand accountability for post-mining management.

The GCB dispute also shows how sampling methods and calculations can influence the representation of the issue. In this case, BRGM opted for a broad range of values, making large deviations appear natural. Conversely, the University of Toulouse geochemists, in the interest of caution for the local population, sought a single average value that was as low as possible. This involved moving away from the contaminated site to sample the wooded areas of the Montagne Noire. While considering the future of life in the area, the researchers were also mindful of its past, before mining and industry had wrought devastation, gathering information by sampling from the strata of the "underground"—its sedimented history.

As a public body tasked with finding a management strategy that is both sustainable and financially viable, BRGM needs to represent the diversity of soil in the region, including areas affected by human activity. The different aims and approaches show their relationship to the future, the past, and space. The scientists qualify the space in relation to a distinct, healthy control area (a reference that is both historical and spatial—"natural"), whereas BRGM composes it from a mesh of points deemed representative, where the past is mixed with the present. We believe that when faced with a matter of delayering, both the historical and spatial dimensions of sociotechnical imaginaries must be considered, beyond Jasanoff and Kim's (2019) conception. This is perhaps the lesson to be learned from this case study, anchored as it is in a specific place where the "turning over of the Earth's crust" puts different, sometimes compressed or mixed, space–time in tension: The past taken as a reference, a future envisaged and desired, and a space into which everyone projects. In these time–spaces, current residents and future generations play different roles.

How can we explain the fact that scientists and engineers, who undoubtedly attended the same universities, proceed in different ways and are unable to agree on results that, as we have seen, are not fundamentally different aside from their presentation? One explanation lies in the framework within which the teams operate. Scientists have access to additional tools that enable them to extend their research further to look for a baseline in relation to soil quality standards. Local BRGM engineers are guided by their organization's national research department, whose directives are not meant for exploration but require strict adherence to procedures for already polluted soil.

However, this explanation seems insufficient. We need to look at the career paths of the individuals concerned. The geochemists we interviewed at Toulouse University were familiar to us. Having encountered some of them during a field study in the Sierra Minera in Cartagena, Spain, we had already expressed our concerns that an overly detached approach could result in sampling and measurement that overlook the needs and interests of local populations (Gramaglia, 2020). Subsequently, they invited us to join them in the Orbiel Valley, where the reservations they had previously shown were no longer viable (notably, they visit this area every year with their students). Residents affected by the 2018 floods asked them to assess the risk of contamination from the sludge. In this tense situation, where authorities were trying to reassure the public, the scientists, whose results were much higher and more alarming than those being circulated, felt compelled to go public. At the same time, they established relationships of trust with certain elected representatives and associations, which facilitated their field access. Thus, a research project was initiated to address the residents' concerns. This approach diverged from the mainstream scientific practices of government departments and expert public bodies, which often distance themselves from, or even avoid, shared experiences. While upholding the standards of their discipline, the geochemists chose to include the social consequences of their research into their methodology. This inclusion has led to a significant reorientation, fostering the development of a different epistemic culture and sociotechnical imagination.

6.7 Conclusion

The residents of the Orbiel Valley were never involved in the technical and economic decisions that paved the way for the rehabilitation of the mining sites. At best, their representatives are now invited to regulatory consultative meetings organized by the prefecture, where monitoring data and information on corrective actions undertaken on waste storage or water treatment facilities are shared. Yet, despite having gained expertise in the field over the years, their questions and proposals are barely considered—at least, not to their satisfaction. Even so, some residents act as sentinels on the ground and have successfully brought malfunctions or issues to the attention of government departments and BRGM. However, the State regards local residents as unreliable sources, given their subjective view of the pollution affecting them, and believes they should be kept far away from risk management. At most, their "perception" of risk is solicited to identify any discrepancies with assessments made by government scientists.

This approach is at odds with the practices developed by the Toulouse scientists, who have fostered a trust-based relationship with several local residents, training them to act as their assistants and partners in the field. These new relationships direct their practices toward greater openness to local social and political issues—beyond economically centric thinking. This echoes developments at some North American mining and industrial sites, where local indigenous populations have been granted a voice in technical decision-making and solutions for post-mining management (Beckett, 2021). Given the intractable nature of the contamination from de-stratification, which will impact several generations, involving local populations is more than welcome. It allows scientific analysis to reconnect with the lived experience of local residents who have witnessed the devastation firsthand. It also opens up discussions on the blind spots in risk assessment, infusing democratic and political significance into the decisions made—toward repairing relationships with the soil and the underground. North American scientific teams were initially reluctant to collaborate with amateurs, victims, or affected groups; they were compelled to engage due to the sovereignty asserted by indigenous communities. This has the advantage of ensuring that issues and disagreements are voiced and not suppressed, as open debate provides opportunities for proposals to be tested and validated collectively. Drawing inspiration from these practices would be wiser and fairer.

Competing Interests and Acknowledgements The qualitative survey from which this article is drawn was financed by the *Occitanie* Region (DIAGNOSE 2020–2023 Project). The authors thank the mayors and residents of the Orbiel Valley, as well as scientists and professionals, who agreed to answer their questions.

Bibliography

Beckett, C. (2021). Beyond remediation: Containing, confronting and caring for the giant mine monster. *Environment and Planning e: Nature and Space, 4*(4), 1389–1412.

Bonincontro, T., Cerceau, J., Tena-Chollet, F., & Becerra, S. (2023). In J. Cerceau & B. Laurent (Eds.), *Quand la mine déborde. Enquêtes sur la fabrique des territoires miniers* (pp. 25–56). Presses de l'Ecole des Mines.

Boudia, S., Reinhardt, C., Creager, A. N., Frickel, S., Henry, E., & Jas, N. (2018). Residues: Rethinking chemical environments. *Engaging Science, Technology and Society, 4*, 165–178.

BRGM. (2021). *Etablissement du fond pédo-géochimique dans la région de l'ancien secteur minier de Salsigne*. Rapport final.

Cerceau, J., & Laurent, B. (2023). Introduction. In J. Cerceau & B. Laurent (Eds.), *Quand la mine déborde. Enquêtes sur la fabrique des territoires miniers.* (pp. 9–21). Presses de l'Ecole des Mines.

Knorr Cetina, K. (1999). *Epistemic cultures: How the sciences make knowledge*. Harvard University Press.

CGPC, CGM. (1998). *Salsigne (Aude)*. Rapport final aux Ministres.

Clark, N. (2014). Geo-politics and the disaster of the anthropocene. *The Sociological Review, 62*, 19–37.

Clark, N., & Rickards, L. (2022). An anthropocene species of trouble? Negative synergies between earth system change and geological destratification. *The Anthropocene Review, 9*(3), 425–442.

Cour des comptes. (2003). *L'Etat face aux enjeux industriels et environnementaux: l'exemple des mines d'or de Salsigne*. Rapport final.

Delplace, G., Viers, J., Schreck, E., Oliva, P., & Behra, P. (2021). Pedo-geochemical background and sediment contamination of metal(loid)s in the old mining-district of Salsigne (Orbiel Valley, France). *Chemosphere, 287*, Article 132111.

Delplace, G., Viers, J., Schreck, E., Oliva, P., & Behra, P. (2022). This letter is a response to the comment submitted to chemosphere by Melleton et al. on our paper (Delplace et al., 2021), published online in Chemosphere in September 2021. *Chemosphere, 307*(Pt 2), 135766.

Gałuszka, A., & Migaszewski, Z. (2011). Geochemical background—An environmental perspective. *Mineralogia, 42*(1), 7–17.

Giddens, A. (1990). *The invention of modernity*. Stanford University Press.

Gramaglia, C. (2020). Saltkrake. Penser la « vitalité » des résidus miniers pour mieux appréhender leurs effets toxiques. *Revue d'anthropologie des connaissances, 14*(14–4).

Granjou, C., Walker, J., & Salazar, J. F. (2017). The politics of anticipation: On knowing and governing environmental futures. *Futures, 92*, 5–11.

Gross, M., & Bleicher, A. (2013). 'It's always dark in front of the pickaxe': Organizing ignorance in the long-term remediation of contaminated land. *Time & Society, 22*(3), 316–334.

Hecht, G. (2023). *Residual governance: How South Africa foretells planetary futures*. Duke University Press.

Jasanoff, S., & Kim, S. H. (Eds.). (2019). *Dreamscapes of modernity: Sociotechnical imaginaries and the fabrication of power*. University of Chicago Press.

Kabisch, S. (2004). Revitalisation chances for communities in post-mining landscapes. *Peckiana, 3*, 87–99.

Kinchy, A. J., Phadke, R., & Smith, J. M. (2018). Engaging the underground: An STS field in formation. *Engaging Science, Technology, and Society, 4*, 22–42.

Le Berre, S., & Chailleux, S. (2021). La relance minière en France et en Europe à l'épreuve des critiques. *Revue Gouvernance, 18*(2), 1–15.

Melleton, J., Belbeze, S., Négrel, P., Brunet, J. F., Vic, G., Auger, P.,... & Laperche, V. (2022). Comment on Pedo-geochemical background and sediment contamination of metal (loid) s in the old mining-district of Salsigne (Orbiel Valley, France) by Gauthier Delplace et al., published in Chemosphere 287 (2022) 132111. *Chemosphere, 307*(Pt 2), 135768.

Mésini, B. (2018). Le laboratoire cévenol de l'après-mine. Une coextensivité des causes et des responsabilités minières, environnementales et sanitaires. *Géocarrefour, 92*(92/2).

Sismondo, S. (2020). Sociotechnical imaginaries: An accidental themed issue. *Social Studies of Science, 50*(4), 505–507.
Ureta, S. (2018). Baselining pollution: Producing 'natural soil' for an environmental risk assessment exercise in Chile. *Journal of Environmental Policy & Planning, 20*(3), 342–355.
Ureta, S., & Flores, P. (2018). Don't wake up the dragon! Monstrous geontologies in a mining waste impoundment. *Environment and Planning d: Society and Space, 36*(6), 1063–1080.

CHAPTER 7

What Can Urban Soils Do? Exploring the Impact of Paris's New Rainwater Management Policy on Soil Production and Maintenance

Maxime Algis

List of Abbreviations

CEREMA	Centre d'études et d'expertise sur les risques, l'environnement, la mobilité et l'aménagement (Center for studies and expertise on risks, the environment, mobility and development)
FAO	Food and Agriculture Organization of the United Nations

M. Algis (✉)
Gustave Eiffel University, Paris, France
e-mail: maxime.algis@univ-eiffel.fr

LISIS, Paris, France

© The Author(s), under exclusive license to Springer Nature Switzerland AG 2025

O. Labussière et al. (eds.), *Back to the Ground*,
https://doi.org/10.1007/978-3-031-88888-5_7

7.1 Introduction: Environmental Framings and the Making of City Soils

Behind the Multifunctionality of Urban Soils: Actors' Conflicts and Competing Framings of the Object "Soil"

This contribution aims to explore the political, economic, and epistemic aspects of the "making of city soils" (Meulemans, 2020) and the role that actors intend for these soils to play in the ecological transition of the urban environment. Informed by observations during meetings between the technical departments of the City of Paris regarding urban soil unsealing policy, these analyses are part of a broader reflection on the emergence of soils as a public environmental problem and how different framings of environmental issues relate to various framings of the "soil" object (Fournil et al., 2018). Each of these framings involves specific discourses on the potential and limitations of soil material transformations.

Among these framings, the interpretation of soils in terms of ecosystem functions and services has gained significant traction over the past 10 years in the discourse of both public and private actors. In particular, it has become a commonly used reference framework in the construction of recent European environmental policies (European Commission, 2023). Following the Millennium Ecosystem Assessment (Reid et al., 2005), this framework aims to link the functioning of soil ecosystems, as described by functional ecology, to quantifiable services rendered to society. The economic valuation of these services would inform decisions based on cost–benefit analyses. Although sometimes depicted as a part of a global campaign to "define the world as an immense collection of service commodities" (Robertson, 2012), the economic approach to ecosystems promoted by this framework appears to be struggling to establish itself in the case of soils (Baveye et al., 2016). However, this discussion will address another aspect of this commodification process: the pervasive idea of "soil multifunctionality" within this framework (Bispo et al., 2016).

This concept of multifunctionality implies that scientific ecology's description of soil as a set of ecological "functions" leads to the notion that soils "naturally" and *simultaneously* assume a variety of important roles (confusion is here maintained between "functions" and "services") for both the environment and human activities reproduction and development. Thus, soil is portrayed as providing habitat for organisms, nutrient cycling, climate regulation, and carbon sequestration while also supporting the provision of food, fiber, and fuel, forming the foundation for human infrastructure, and facilitating water purification and soil contaminant reduction. The Food and Agriculture Organization of the United Nations's "soil functions" diagram with the subtitle "soils deliver ecosystem services that enable life on Earth" exemplifies this perspective (FAO, 2015). However, this representation, by equating these varied functions or services within the same "multifunctionality" concept, glosses over any potential forms of incompatibility or conflict among them. For instance, extracting construction material, as done by quarrying companies, results in soils incompatible with their function as habitat for organisms—although ironically also presents opportunities for soil ecologists to propose restoration and reconstruction programs (Menta et al., 2014). On another topic, the ability of soils to regulate climatic events often clashes with how industrial agriculture utilizes soil for food production (Lin et al., 2011). Therefore, this representation tends to obscure the contentious aspects of soil management and protection by overlooking political (societal choices regarding modes of production, conflicts over the collective value of specific soil uses), economic, and technical stakes (the contradiction between how a given "service" transforms, and sometimes degrades, the soil functions that would support another service's sustainability). This approach is also increasingly applied to urban soils, where some stakeholders aim to "build decision support systems that take into account the ecosystem services provided by soils" to develop sustainable management of urban areas, focusing on "optimizing the ecosystem services provided by urban soils" (Blanchart et al., 2018).

In contrast to these views and their normative ambitions, this contribution does not seek to describe and quantify the functions and services that urban soils could render to "society." Instead, it aims to describe how various social actors (representing here different departments of the City of Paris) construct both the uses and the materiality of urban soils. This conflictual process is intended to negotiate what urban soils are and could be capable of "doing," based on political, economic, and technical considerations. These technical considerations, while incorporating descriptions derived from functional ecology, are also influenced by the institutional (traditional organization of professional skills in the City's departments) and sociotechnical (hygienist urban operating model inherited from the nineteenth century) structuration of the environment in which the actors find themselves.

This contribution aims to highlight the diversity of relationships to urban soils among these different departments, as well as the tensions and contradictions made visible by the project of a "pervious city" and soil unsealing, which tests the established division between these departments for the production and management of urban soils.

Enrolling Technical Departments in the Unsealing of Urban Soils: The City of Paris's "Rain Plan"

The project to restore the city's "natural water cycle"[1] by improving rainwater management is typically considered part of the transition toward a "sustainable city." In this sense, reopening urban soils to allow rainwater infiltration is presented as a means to redeploy nature in the city, reactivate interactions between plants, soils, and rainwater, and grow urban biodiversity.

However, the ambition for "alternative" rainwater management in cities (or "sustainable drainage systems"), which emerged in the 1970s and refers to management that would allow less of this water to be sent to the sewer system, originally had little to do with constructing a new "renaturalized" urban landscape or reconstructing urban soils' ecological functions. It was initially a response to infrastructural problems related to the management of urban drainage systems, which were often overloaded by rainwater (Barles & Thébault, 2018). By ensuring that rainwater does not reach the sewers, the objective of sanitation services is to avoid flooding and contamination of rivers by preventing the network from overflowing. Thus, paradoxically, the actors originally promoting the project of a pervious city and framing the sealing of urban soils as a problem to be solved are historically those who are relatively blind to the functioning of the urban surface (they are in charge of the management of underground networks) and whose existence was originally conditioned by the hygienist project of soil sealing. From the nineteenth-century onward, this project has articulated the pipe underground and the asphalt on the surface in a complementary manner (Barles, 1999; Zitouni, 2013).

[1] Also known as the "large water cycle," in contrast to the "small water cycle" (or urban water cycle). This latter describes a cycle that begins with pumping water from a river, followed by potabilization, use, and then sanitation of this wastewater, with its discharge back into the river. Rainwater is incorporated into this cycle as it is collected in the city's sanitation systems in the same way as wastewater.

This paradoxical position leads urban sanitation technical departments to translate their infrastructural problem into an "urban issue" in the broad sense. By collaborating more closely with planning actors and ecologizing the rainwater issue, the mainly technical solutions of the early days are gradually being replaced in their discourse by the project of an "upstream" rainwater management (to infiltrate water where the raindrop falls on the surface). This option is presented as an opportunity to exploit rainwater as a resource in newly re-created ecosystems that would contribute to a renewal of urban nature. Aware of the importance of these ecological translations in enrolling other urban actors, the City of Paris's sanitation department[2] is using the full range of arguments offered by the environmental issue, including the concept of soil multifunctionality improvement. Its goal is not only to convince the city's stakeholders in the broadest sense but also to campaign within the municipal institution to promote its objectives among elected officials and other technical departments. Thus, stormwater management is presented by the sanitation department, not as solving a technical problem related to urban sanitation but rather as "a small revolution in the way of conceiving the city and a collective challenge [...] for the benefit of all" (Mairie de Paris, 2019) and as a means of building a "sustainable" city. In this context, urban soil transformation is strategically positioned as an opportunity to enable the provision of new "ecosystem services." The climate issue is engaged through the issue of urban heat islands—presented as a consequence of soil sealing—while unsealed soils are touted as the most cost-effective solution for managing both chronic and accidental pollution (Nezeys et al., 2016). The issue of water as a limited resource is also highlighted: "water is saved because rainwater is recovered for uses that do not require drinking water (sanitary facilities, watering, cleaning, etc.)" (Mairie de Paris, 2019). Moreover, biodiversity is expected to "flourish" under this new rainwater management policy.

The rise of the sustainable development notion, which became a central reference for public planning policies in Europe around the turn of the century (Emelianoff, 2007), may explain the success of this greening strategy. In Paris, this led to the political decision in 2018 to adopt a "Rain Plan," which, as argued by its creators (i.e., engineers from the sanitation department and environmental consultancy hired by this same

[2] Officially called STEA for Service Technique de l'Eau et de l'Assainissement.

department), is "probably the first in France to have taken purely environmental objectives into account in its prescriptions," including "in particular, ecosystem benefits that are not necessarily hydraulic, such as the fight against urban heat islands." Thus, rainwater zoning, a regulatory component of the Rain Plan, "becomes a component of the city's overall environmental policy, of which it integrates related issues" (Nezeys et al., 2016).

But if these "related" issues are indeed considered, and if this plan is firmly aligned with the city's "global environmental policy," how can we account for the fact that other departments of the city, also committed to this "sustainable city" objective, appear to struggle with embracing the alternative stormwater management approach? How can we explain these reluctances, such as those of the urban planning department,[3] which refuses to participate in the sanitation department's meetings on the Rain Plan, even though two years earlier it had published a revision of its local urban plan introducing "new or enriched environmental provisions to better take into account this ecological transition" (Mairie de Paris, Direction de l'Urbanisme, 2018)? Or those of the parks department,[4] which has been working on a Paris Biodiversity Plan, a document that only marginally addresses the issue of rainwater, since 2015?

From Different Epistemic Cultures and Their Relationships to Soils to Deeper Concerns About the City's Geological Strata

The observations on which this contribution is based allow us to explore the gap between, on the one hand, the display of a consensual project of transitioning toward a "sustainable city"—supposed to involve, among other things, urban soil unsealing and the development of new ecosystem services by the newly unsealed soils; and on the other hand, the conflictual nature of technical department negotiations concerning the acceptable and non-acceptable conceptions of what urban soils can and must do, what can be asked of them, and how they can be transformed. These negotiations about the definitions of what makes "good" or "bad" soils—what threatens them or what can sustain them—are more than just

[3] Officially called DU for Direction de l'Urbanisme.

[4] Officially called DEVE for Direction des Espaces Verts et de l'Environnement.

questions of technical feasibility. By challenging the prerogatives and practices of various professions organized in each city's department, the issue of rainwater infiltration makes it necessary for these actors to clarify their diverse and sometimes contradictory relationships to the object "soil."

Each city's department, in fact, possesses its own "epistemic culture," defined as "sets of practices, arrangements, and mechanisms bound together by necessity, affinity and historical coincidence which, in a given area of professional expertise, make up how we know what we know" (Knorr-Cetina, 2007). In each of these different "self-referential systems that orient more to internal and previous system states than to the outside environment," various "specific constructions of the referent" (Knorr-Cetina, 2007)—here, of the object "soil"—were developed, closely linked with the professional practices (Nerland & Jensen, 2012) of each "community of technological practitioners" (Constant II, 1987) associated with each city department. These differentiated relationships with the object "soil" have been shaped within the framework of an urban planning tradition that has defined specific forms of work organization in the structure of the municipal administration. Depending on the department and the professional practice, these different epistemic cultures carry with them articulated sets of technical and historical references, values, habits, gestures, affects, and relationships with objects that coexist but are rarely brought into contact with each other outside of the institutional routines that compartmentalize the fields of intervention of each—fields whose boundaries are, in our case, blurred by the transversal problem of infiltration.

As a result, the first part of this contribution examines the different epistemic cultures of each city department and the misunderstandings and divisions that arise in discussions on rainwater management at the city's surface. These divisions are conditioned by the differing relationships each community of practitioners has constructed with the object "soil." In the second part, we will take a step back from analyzing these incompatibilities between city departments to explore what, in the process of making soils, unites them: shared views of soil as a project matter and as a maintenance problem. We will then investigate how, despite these opportunities for agreement in the construction and maintenance of new permeable soils, several concerns persist regarding the risks posed by rainwater infiltration to both the anthropic and geological organization of urban subsoils, revealing the interdependence between the policies that aim to govern these different strata.

7.2 Plural Epistemic Cultures and Relationships with Urban Soils: The Pervious City's Materiality Challenged

City Department Confrontations

The research for this contribution was conducted during meetings organized to draft a guide supporting the implementation of the new Rain Plan developed by the sanitation department. These meetings, hosted by the sanitation department, brought together various departments of the City of Paris to co-define recommendations for implementing this new regulation, primarily aimed at private actors. However, a second objective of these meetings was to align the design, construction, and management practices of the other departments with the objectives of the sanitation department's Rain Plan.

Participants in these meetings included representatives from the sanitation department (both management and technical teams), as well as from the roads department,[5] parks department, quarry service,[6] building permit department, urban cleaning service,[7] housing department,[8] and the youth and sports department,[9] along with actors who participated occasionally depending on the topic (developers, social housing providers). An engineer from an environmental consultancy, hired by the sanitation department to author the guide and facilitate the meetings, was also present, alongside representatives from the Parisian urban planning workshop[10] to discuss urban planning issues and assist with the guide.

My attendance at these meetings (seven meetings lasting about 3 hours each, including a preparatory session with the environmental consultancy) stemmed from my role as a research officer at this urban planning institution at the time. I was invited to assist a senior colleague and prepare reports, providing a unique vantage point to observe the positions of all stakeholders. During this period, I had not yet been exposed to social science methods of inquiry. It was only a year later, through

[5] Officially called DVD for Direction de la Voirie et des Déplacements.

[6] Officially called IGC for Inspection Générale des Carrières.

[7] Officially called STPP for Service Technique de la Propreté de Paris.

[8] Officially called DLH for Direction du Logement et de l'Habitat.

[9] Officially called DJS for Direction de la Jeunesse et des Sports.

[10] Officially called APUR for Atelier Parisien d'Urbanisme.

personal notes, reconstructing exchanges ex-post, and retrieving e-mails and certain meeting minutes, that the present analyses took shape. My method involved first reconstructing the chronology of meetings and participants, recalling the exchanges actor by actor, then articulating the observed positions, staging their tensions and agreements, and finally, comparing these with other sources, such as the guide ultimately produced and gray literature from the City of Paris.

The discussions observed provided insights into the reluctance shown by actors responsible for roads, parks, or former quarries. Their differing conceptions of urban soils, as matters of design and maintenance, vary according to their epistemic culture and their role in the organization of urban services. This diversity of relationships with soil conflicts with the proposals of the sanitation department, which, historically lacking a specific urban soil culture beyond that of infrastructure, promotes stereotypical technical devices typically developed by hydraulic engineers in the 1980s and 1990s. If the discourse on rainwater management has progressively been framed as an urban environmental issue since the 1970s, "The range of techniques developed at that time varies very little until today: open or buried retention basins, ditches [...] drainage trenches, infiltration wells, reservoir or vegetated roofs, porous pavements" (Barles & Thébault, 2018). This inventory aligns almost perfectly with the devices that the sanitation department initially intended to include in its guidebook.

It can be assumed that the sanitation department, aware of the controversies over the actual implementation of the measures it advocates, preferred to address this "technical" issue at the end of the process (by the time of these meetings, the Rain Plan had already been approved by the elected officials) and in a guide separate from the regulatory documents. With this strategy, and by emphasizing global urban ecological issues, the sanitation department tried to maintain control over the rainwater infiltration issue as long as possible, even though this challenges the "silo-like" organization of the technical departments responsible for urban soil. Thus, the meetings I attended became the unexpected venue for a "return of the repressed," where other departments seized the opportunity to voice their dissent and assert their own epistemic culture and relationships to urban soils. Each request from the sanitation department for validation of the solutions presented in the draft guide sparked a tense debate about what the city's soils are and should be, according to each of the actors present.

In fact, each representative embodies a distinct "soil voice" that often contradicts the other positions expressed. Although the distribution of action perimeters and objectives among these services has historically been aligned with the same hygienist city project, the outputs of each department being complementary (the mineral pavement laid by the roads department channels surface water to the underground networks of the sanitation department, and the soil managed by the parks department supports vegetation that helps "aerate" a predominantly mineral urban environment). Yet, their interventions on urban soils occur in adjacent compartments and are not easily superimposed. Thus, each department representative defends their professional practices in interpreting the new rules and devices proposed by the sanitation department. The parks and roads departments are the most active participants in these discussions; each represents a range of skills, and expertise that they are willing to contribute to the municipal project, provided these are recognized and considered. Here, the imperatives of rain management clash with the concerns of each profession: For the parks department, these include risks of over-watering, soil pollution, and the leaching of hydrocarbons into planted spaces. For the roads department, concerns are organized around the solidity of the road structures and the risks of damage, as well as the difficulty of integrating "infiltrating" solutions into a repertoire of rigorous and codified work practices. This reticence clashes with the demands of the sanitation department, which views urban soils primarily as a technical layer through which its networks operate.

Sanitation Department and Soil as a Technical Layer Disconnected from the Uses of the City's Surface

Representatives from the sanitation department express a soil culture where soil is seen as a technical layer dedicated to the circulation of flows (drinking water, wastewater drainage). The crisis of this culture, viewing soil as a space for circulation and infrastructure storage, is at the root of the disputes observed here and of the request made to other actors to reconfigure their relationship with the soil.

This crisis is linked simultaneously to the increasing demands regarding the quality of water in rivers, where overflow from sewer systems during rainy episodes is the primary factor of pollution, and to a growing economic challenge in responding to the saturation of this network with solutions like large underground infrastructures (reservoirs or stormwater basins). In the case of Paris, this ecological demand is coupled with an urgent political will related to the organization of the 2024 Olympic Games and the hosting of sports events in the Seine.

Thus, while the sanitation department never completely abandons its discourse on sustainable cities, it nevertheless presents the transformation of existing public spaces as an urgent technical necessity and a call for participation by all the city departments in addressing sanitation issues. This stance, which overlooks the views, constraints, and timelines related to the work of other departments, is frequently the cause of breakdowns in communication. The sanitation department's demands that other departments take rainwater into account are based not only on the political legitimacy provided by its greening efforts but also on an evolution of public policies that have been taking place at the national level since 2010. The new regulatory possibilities[11] have been used by this department to enact rules while bypassing any collaboration with the urban planning department to develop an urban strategy for rainwater management. This "avoidance" and encroachment on the regulatory power of the urban planning department is perceived very negatively, as evidenced by the absence of representatives from this department at the meetings.

Urban subsoils continue to be viewed by the sanitation department as an essentially technical system, the most pressing issue being the reduction of the volume of rainwater—considered one of the inputs to this system—entering it. The potential problematic consequences of urban soil unsealing and rainfall infiltration are, therefore, recognized when they relate to soil stability (risk of collapse) since they jeopardize the state of the networks, but they become more difficult for these actors to understand when expressed in terms of the city's surface use distribution.

Parks and Gardens: Open Soil and Management of the Living Environment

Representatives from the parks department show some mistrust toward the sanitation department, which advocates for managing the entirety of

[11] Since 2010, it has been possible to annex to local urban plans, in addition to the classic sanitation zoning, a pluvial zoning that designates areas "to limit soil sealing and to ensure the control of the flow and runoff of rainwater and run-off water" (Loi n° 2010-788 du 12 juillet 2010, 2010).

rainwater at a local level and outside the network. Indeed, beyond the unsealing of existing mineral spaces, the refusal to accept rainwater in the sewage system implies managing it on the surface, which includes infiltrating it into spaces already in open ground. Officials from the parks department, therefore, fear this new policy will lead to the saturation of the spaces for which they are responsible, by redirecting the collected rainwater to all the surrounding green spaces.

The initial perception of soil expressed by these actors is that of a "natural" soil, considered a rare resource in a dense, mineral Parisian urban fabric. The concern is that concentrating constraints on these rare spaces could compromise their "natural" infiltration capacity, endangering the soil's physical structure and surface condition, and potentially causing problems such as water stagnation and mud formation. This refusal is based on the conception of soil as a fragile system in equilibrium and on the parks department's mission to maintain its stability.

The second perception of soils put forward by the parks department representatives is that of soil as a growth environment, a living soil. While the stability mentioned above refers to the physical components of the soil, the balance here is dynamic: It includes non-human animal and plant populations that circulate, grow, and transform. The intent to modify the quantity of water in these soils poses the risk of unbalancing much of the plantations. Besides over-watering, the main risk highlighted by these actors is the toxic threat that potential pollution poses to living soils: Rainwater will transport various elements from the urban space to the "natural" soils, whose "good management" until now has meant keeping them as isolated as possible from the rest of the urban surfaces. Metals from roofs (zinc), oils, and heavy metals from roadways, cleaning products, and event-specific products (snow removal salts, hydrocarbons from road accidents) pose numerous risks to living organisms. In this context, the parks department is striving to curb the tendency of other departments to embrace the concept of phytodepuration to prevent planted areas from being progressively and systematically considered as pollution filters for runoff from other urban surfaces.

Although the representatives of the parks department discuss "natural" and "living" soils, they acknowledge their constructed aspect: Plant varieties are carefully selected, substrates are studied to optimize growth and resistance, and the introduction of new species or processes (mulching, composting, etc.) is undertaken with caution. The discourse on the fragility of living entities complements a discourse on investments in terms of labor and expertise. Another "human" element that paradoxically also defines these "natural" soils in the parks department discourse is that of usage and users. The new role claimed by the sanitation department for parks and gardens, that of "buffer zones" for rainwater infiltration, could threaten their primary function as spaces for leisure and walks. Accessibility issues are then raised as obstacles to rainwater management. If these spaces receive a large concentration of water and become wetlands or ponds, they will need to be protected and will exclude the public. The influx of water here threatens the hygienist definition of nature that associates it with health. Thus, in defending the public use of green spaces considered urban sanctuaries, the hygienist ghosts of miasma and stagnant water, of mephitic sludge, and health risks (Barles, 1999) (which here manifest as uncertain risks linked to potential rainwater pollution) resurface.

Roads Department: Soil Viewed as Structure and Surfacing

The roads department logically shares this last conception of urban soils as constructions. For representatives of this department, soil is primarily an object of design adapted to the needs of surface uses: the hierarchy of lanes and the frequency of traffic. However, unlike the parks department, which designs an ecosystem functioning as an interface between flora and the mineral subsoil, the roads department focuses on designing a stable structure in the upper part of the soil to isolate the surface from the lower layers. The two main concepts evoked by representatives of this department are, therefore, structure and surfacing.

The design of the structure involves a judicious choice of the different layers that compose the ground. Surfacing refers to the final surface layer, used for vehicular or pedestrian traffic. It is interesting to note that in the context of road construction, the term "profile" no longer refers to the "thick" soil profile of the pedologist showing the different soil horizons but to the outline of the last and thin surface layer. Thus, the profile here refers to the design of a "cross-section" that represents the inflections and indentations of the pavement from one side of the street to the other, allowing for both the functional decomposition (pedestrians, vehicles, parking) and the path of rainwater from the sloping surfaces of the road and the sidewalk to the gutters and the sewers. In these meetings, the sanitation department aims to persuade representatives of the roads department to consider these two conceptions of the "profile" simultaneously, linking the department's historical expertise in surface water management with the project of controlled diffusion of rainwater within the thicknesses of the soil and the subsoil.

However, for the roads department representatives in attendance, the concepts of permeable sidewalks or pavements are not very convincing. These ideas contradict their structured vision of soils: The surface profile is designed precisely to allow water to be "chased" toward the networks, thus preserving the structure of the roadway and its organization. In this context, soil porosity is viewed as a defect, literally a fault. Therefore, soil unsealing is taboo for the roads department, seen as a step backward in the history of urban infrastructures. Yet, it is this very notion of history that will form the basis of the discussions. The stance of the roads department's representatives is ambiguous and uses two different registers of criticism against the idea of permeable soils, varying by the occasion (and personalities involved). The first sees porous pavements as a step backward when techniques are constantly improving; this "progressive" register uses the image of medieval pavements as a deterrent. By contrast, the second, more "nostalgic" register criticizes porous pavements with the opposite arguments: The new "innovative" permeable pavements are seen as contradicting the time-honored rules of pavement laying mastered by the professionals of the roads department.

By advocating for the preservation of traditional knowledge, certain participants (from the sanitation department and the Paris urban planning workshop) attempt to present the reintroduction of water to the surface and subsoil as an opportunity to revalorize skills linked to roadwork professions, such as the increasingly rare art of paving. For the sanitation department, paved surfaces could be porous if the joints were filled with sand, allowing rainwater to infiltrate in a distributed manner. These proposals, found appealing by roads department representatives who value the recognition of their profession, quickly face the realities of work organization on road construction sites. The roads department now heavily relies on service providers and performs minimal work with its own teams. Road projects are designed by the City but are then executed by actors whose skills align with the most widespread and "industrializable" techniques: gravel and concrete base layers followed by the application of a liquid surfacing layer of bitumen or asphalt.

The gathering of representatives from departments managing the city's underground networks, green spaces, and roads facilitates an understanding of the functional decomposition from which urban soils and subsoils are considered in the organization of the city's services. The absence of members from the urban planning department further underlines how soils are missed in a cross-cutting analysis here and are generally considered from the kaleidoscopic multiplicity reflected by the various departments organized in silos, each with its own epistemic culture.

7.3 From (Future) Urban Permeable Soils to Geological Concerns

Imagining New Soils

Despite the reluctance of the departments mentioned earlier, these meetings represent an opportunity for each department to demonstrate its goodwill in terms of ecological transition and, thus, rainwater management. This is generally done by highlighting a completed or ongoing project: sections of streets with grassed paving stones, gardens that collect water from the roofs of neighboring buildings, etc. However, it appears that these experiments generally represent exceptions: In one specific case, the technical department concerned invested resources in time and human material to implement a "model" device, which was rarely followed by a generalization of this device to other projects within the department. In contrast to these one-off projects, we will see that during the meetings, some points of discussion suggest possibilities for thinking about urban soil design methods without having to resort to the stereotypical "alternative technical solutions" typically proposed in the technical guides written by the sanitation actors.

We have already observed that "natural" soils in cities are presented by representatives of the parks department as the result of construction and maintenance efforts. Building on this concept of "constructed" soil, an engineer from the city's agronomy laboratory suggested during a meeting a possible adaptation to the new challenge that represents a greater concentration of rainwater in planted areas. Other departments' representatives, recognizing aspects of their own daily work, engage in this process of envisioning potential structures and functions of the soil, leading to an exchange of ideas about the possibilities of constructing these new soils in various ways. Thus, challenging the automatic recourse to imagery of wetlands or ponds—often associated with strong water supply—this agronomist from the parks department introduces the concept of constructing dry soils with associated flora, designed to cope with both massive and irregular water supplies, taking some Mediterranean soils as a model. This proposal shifts away from the classic images of ditches and other seepage areas, supposedly stabilized around a new "wet" equilibrium, by replacing them with the image of a soil capable of rapidly migrating humidity to its lower layers. This example helps distance the hygienist's fear of stagnation and imagine new urban soils using materials and functions familiar to the technical services.

Although this moment of discussion is brief compared with the presentation of technical impossibilities by each city department, it nevertheless informs us about the potential role of imagining soil complexes and their functioning in a cross-departmental approach to soil management. Once the present actors are freed from the need to assert their rejection of predefined technical solutions, they readily pool their technical knowledge and their capacity for innovation and adaptation around the design of a system likely to respond to the constraints of each service: solidity, human uses and circulation, vegetal growth, and vertical infiltration of water.

The Collective Problem of Maintenance: the Key to a Common Culture of Urban Soils?

If, with the implementation of a sort of "pedological imagination" as mentioned above, the present actors see the possibility of collaborative work in the short-term process of designing soils differently, the question of maintaining infiltration facilities continues to bring these same actors back to the challenges of long-term management through repeated shared practices (Denis & Pontille, 2020). However, it seems that, as a problem shared by each department, maintenance could also provide fertile ground for the collective invention of new forms of urban soil management.

The issue of maintenance makes it necessary to discuss the role of the urban cleaning service. During the meetings, representatives of this service show little interest in fueling controversy, even though they are regularly called upon by various participants to confirm the difficulties of managing and maintaining rainwater infiltration facilities. The urban cleaning service fundamentally views its role as complementary to that of the sanitation department: to wash urban surfaces in order to push waste toward the sewage networks. But if rainwater is to be considered a resource that can be collected, used to water green spaces, or returned to the lower layers of the subsoil, how can waste be removed from these new specific flows? At the same time, the conceptions of clean and dirty, the grammar of urban spaces and their surfaces, and the working methods of maintenance teams are called into question by the "upstream" management of rainwater. This issue of maintenance arises during discussions about the technical devices to be implemented for in-situ infiltration, evaporation, or reuse of rainwater. These devices are largely "hybrid" in nature, encroaching on the competencies of different departments: infiltrating roadways, planted ditches alongside roadways or sidewalks, planted and lowered tree bases that recover and "buffer" rainwater from the sidewalk, mineral or planted "flooding" zones in public spaces, paving stones, and infiltrating pavements, underground storage tanks for irrigation, etc. These hybrid objects, presented as "innovative" by the sanitation department, alarm other department representatives because they deviate from the usual classifications of urban facilities. Their hybrid nature makes sharing maintenance tasks between departments nearly impossible, fueling a perpetual blame game: An infiltrating roadway can be considered both a roadway and a sanitation facility. Sections of sidewalks, tree bases, or grassed and planted ditches fall under the responsibilities of both the parks and the roads departments, and depending on the complexity of the systems (presence of an oil separator, small retention basins), also the sanitation department. The overlap and specific timing of tasks such as trimming, mowing, cleaning, maintaining technical equipment, and verifying the proper functioning of the structure advocate for new forms of maintenance, yet as it stands, it perpetuates a trench war where each party defends its expertise. The economic question is also at play: The increase in surfaces needing maintenance does not necessarily come with an increase in financial resources for the departments, so each department seeks to minimize the stock of structures for which it is responsible.

It is worth noting that the urban cleaning service is marginally considered in this debate. Unable to mobilize "valued" expertise, this service is alternately cited by the other parties as an obstacle: Its current cleaning practices (broom trucks, pressurized water jets) will struggle with the proliferation of grassy structures, pavements with infiltrating joints, and the reappearance of natural soils, mud, and puddles. The increase in its workload and the prospect of an army of sweepers forced to manually collect waste trapped in the tall grass are repeatedly brandished as a scarecrow by representatives of the roads and the parks departments.

How all the technical questions specific to each department and profession converge around the question of maintenance suggests this may be the key point from which it might be relevant, as sociologists Jérôme Denis and David Pontille (2022) have done in other fields, to re-examine the actors whose "postures" regarding urban soils were described above. These actors naturally and immediately value the forms and practices linking their professions to the construction or production of soils but generally leave in the shadows, unless specifically tasked with a particular installation, the concerns each service has with the daily and repeated procedures of maintenance.

Deeper Concerns. Transforming Urban Soils, Endangering Subsoils? Vertical Interdependency and Politics of Strata

In the points of agreement or confrontation that emerged during the discussions, a potential "common ground" appears on which the city's various departments could begin to elaborate a common definition of soils and their capacity to absorb rainwater. However, this potential consensus did not alleviate other concerns expressed during the discussions, which persisted right up to the end of the meeting cycle: the risk of destabilizing urban subsoils through the success, or at least the progress, of soils unsealing, and therefore the increased volume of rainwater infiltrated into the subsoil. These concerns are echoed by the quarry service with its warnings about the connection between water infiltration and the risk of subsoil collapse.

Here, two relationships to the "*surplus* of potentiality exhibited by the Earth" (Clark, 2017) (the soil stratum, as well as the geological or the atmospheric one) are in competition: One involves actors grappling with overflowing sewage systems triggered by the randomness of rainfall episodes; the other concerns actors dealing with the potential for melting and sliding of underground mineral strata. The former seeks to refabricate the city in a different relationship to rainwater by reauthorizing it to cross the soil strata. However, "putting this surplus to work remains risky too because the environments into which new techniques, productions, and assemblages are introduced are themselves inconstant—reinventing their own criteria of acceptability as they go along" (Clark,

2017). For actors working with geological strata, these transformations will necessarily impact the expression of the "surplus of potentiality" in urban subsoils and the fragile balance of existing assemblages. The "surplus" water that urban soil actors are seeking to redirect is actually not disappearing but would potentially circulate differently and be stored elsewhere, potentially in Parisian aquifers, thus revealing other forms of compromise between human activities and "natural" processes.

In this sense, the quarry service representatives at these meetings reflect one of the current policies for governing the urban underground strata: that of controlling and maintaining their solidity by preserving a state of "stasis." This conservation also includes preserving a history of underground extraction through the maintenance of archives: cartographic knowledge that enables monitoring of the city's underground thicknesses. These quarry services' "conservation" concerns align with those of other actors involved in the anthropic life of the subsoil in their fight against groundwater. For the quarry service, rainwater represents a chemical agent capable of dissolving certain rocks in the infiltration path. For those involved in maintaining and developing transport networks, the tunnels they dig, equip, and use intersect with Parisian aquifers, which threaten them with flooding. Infiltrating rainwater means adding volume to the groundwater against which they are constantly battling.

Thus, representatives of the quarry service were present with their extensive knowledge of urban underground during the development of the rainwater zoning (the Rain Plan) and in meetings, participating in drafting restrictions to limit rainwater infiltration in specific areas. During the debates we observed, the quarry service consistently highlighted the potential dangers in areas of former quarries, gypsum, swelling clays, or poor-quality backfill. It regularly emphasized that in the absence of sufficient feedback and consequent quantified studies, the precautionary principle must prevail against risks that are both geographically identifiable and uncertain. The superimposition and addition of these risks result in a map of Paris where approximately 80% of the surface is vulnerable, placing the discordant "voice of the subsoils" fundamentally in contradiction with the objectives of the sanitation department. Consequently, warnings from the quarry service are a recurring motif during the meetings, which, because of its frequency, is no longer really heeded by the other interlocutors: The focus on tensions between the roads, parks, and sanitation departments over soil definition and deconstruction makes these alerts unrecognized here.

Therefore, a new dividing line is emerging, different from those identified between the sanitation, parks, and roads departments: an opposition between actors linked to soil work (including the managers of the sanitation network) and those linked to the subsoil. As mentioned before, alongside the quarry service, we could also include the managers of the subway network, who were absent from the meetings but whose concerns about tunnel flooding are regularly cited against rainwater infiltration projects. These underground actors are confronted with the potential consequences of the unsealing policy: first, the disorganization of the mineral layers poses risks, particularly in areas with Ludian gypsum, of gypsum compartments dissolving, which can lead to underground collapses and even sinkholes, or in areas with clay, where swelling and then shrinkage of certain subsoil layers could result in structural damage. In addition, there is the potential rise of the Parisian aquifers, which are already artificially maintained below their natural level by permanent pumping in certain areas of the capital. In the model of a pervious city, the subsoil becomes the primary recipient of rainwater, replacing rivers currently situated at the "end of the pipe". This new status raises concerns among the actors responsible for the current uses of these subsoils. It is evident from the position taken by the quarry service that they find it extremely challenging to align themselves with the project of rainwater infiltration.

Differentiated policies are becoming apparent here, depending on the strata under consideration: on the one hand, the crisis in sanitation infrastructures and an urban soil ecology that promises a sustainable city through de-permeabilization, and on the other hand, a policy of managing the risks associated with subsoil destabilization. Certain actors at other scales of regulation, who are seeking to build new forms of knowledge that will enable them to govern these strata together, acknowledge and try to reduce this antagonism. In the regional context, actors like CEREMA[12]—a public institution providing expertise on environmental and risk matters—have established a working group to develop a body of knowledge that is not only cartographic but also pedological. This allows for assessing the actual risks according to the volume of rainfall infiltrated and the type of soil (Dumont et al., 2020). The aim is to establish quantifiable data on the response of a wide range of "at-risk"

[12] Centre d'études et d'expertise sur les risques, l'environnement, la mobilité et l'aménagement.

urban soils to infiltration tests, in order to develop classifications based not on hazard but on the soil's potential in terms of the concentration of water to be infiltrated.

7.4 Conclusion

The debates among practitioners analyzed here have helped us move beyond an approach to urban soils based solely on their potential "multifunctionality"—a framework that often runs risks conflating the description of soils proposed by functional ecology with a functionalist interpretation of the social relationships to soils and their materiality. Paying close attention to the discourse and conflicts among the various actors responsible for maintaining and transforming urban soils for the City of Paris, it becomes clear that the diversity of epistemic cultures associated with each community of technological practitioners plays a central role in negotiating the future of urban soils. In this context, urban soils cannot fulfill all the functions they are theoretically asked to perform simultaneously within this concept of multifunctionality. The specific perspectives of each actor group mean these different functions are considered from cultural, economic, and political angles, and it is through the negotiation of these representations that the compatibilities and incompatibilities between different soil functions and material realities are gradually constructed. Thus, the discussions forming the basis of this contribution highlight the importance of organizational, economic, historical, and political factors in the sociotechnical controversies surrounding the creation of urban soils and how contemporary ecological issues are likely to challenge the institutionalized practices of urban space production.

The ambition to manage rainwater at the city's surface is unique in that it concerns a matter claimed by several different technical departments. Urban soils are indeed a central issue for the roads department, the parks department, the urban cleaning service, and the urban planning department. The demands made by the sanitation department for a global transformation of the functioning and materiality of these soils (to prevent the sanitation network from becoming saturated by rainwater) challenge the existing organization of the production and maintenance of urban soils. This "silo-like" organization, inherited from the spatial and administrative design of a hygienist city, remains viable only as long as each function assigned to urban soils by each department is independently produced and managed. However, new motifs of infiltration, redirection, open-air storage, and evaporation of rainwater are blurring the established boundaries between the categories in use. Thus, behind the declared intention of each actor to participate in the municipality's ecological project, negotiations are taking place among departments regarding the new acceptable definitions of what urban soils can and should do, the distribution of responsibilities for their production and maintenance, the economic implications of these new competencies, and, above all, the practices and technical categories that can be collectively referenced.

As shown, observing these negotiations provided an excellent opportunity to describe the categories and practices related to soil that each of these department's values, as during these discussions, representatives from these different "epistemic cultures" each sought to assert the legitimacy of their approach. But beyond highlighting the fault lines and points of controversy that divide these different relationships to the object "soil," this contribution also aimed to explore the ambivalences of a potential collective design of urban soils. Some exchanges have revealed, beyond the firm positions taken by the departments, calls for a reorganization of the roles and categories used in the production of these soils. These calls were seen by the actors as a prerequisite for a real change in practices. During these moments of collaboration, the actors demonstrated a capacity to "imagine" new forms of urban soils that could functionally meet the new demands of the sanitation department while also valuing the skills and technical traditions of the various departments and uniting these professionals around new metaphors, such as that of the "Mediterranean soil."

While possible paths are being developed for a common conception of new forms of urban soils, the question of maintenance continues to impede progress in the debates. Indeed, although the hybrid nature of the new systems proposed for managing rainwater on the surface (e.g., infiltration channels, porous pavements, and lagoon areas) seems acceptable to the technical departments during the design phase (through greater investment in resources), it inevitably becomes problematic during maintenance. The overlapping maintenance tasks between each department are controversial. One perspective of this contribution is to consider that if the problem of maintenance crystallizes much of the discourse on the technical impossibility of creating a pervious city, it also potentially represents the point at which new "transversal" and inter-service ways of conceiving urban space and organizing its management could be developed.

But beyond these paths toward constructing a common urban soil culture among the various practitioners, disciplines, and departments of the City of Paris, other concerns emerged during the discussions that no longer directly concern the materiality and functioning of the soil strata but that of urban subsoils. The representative from the quarry service spoke on behalf of the actors involved in the anthropic and geological functioning of these subsoils. For the policies aimed at stabilizing this stratum, the infiltration of rainwater is considered a danger, as these new volumes of water risk destabilizing subsoils and flooding infrastructures. This discordant voice highlights not so much a confrontation between different epistemic cultures that have developed sometimes opposing relationships to urban soils but rather a confrontation between different politics of strata. Faced with efforts to govern the circulation of water in the city's soils differently, we encounter other actors with another structured relationship to "the matter, the flows, the reservoirs." These actors fear that their own ability to govern the functioning of the subsoil will be surpassed by "deeper forces and formations that will stubbornly cleave to their own agenda" (Clark, 2017): the dissolution of minerals and the rising of groundwater.

Competing Interests and Acknowledgements The author has no competing interest to declare. He warmly thanks Frédéric Bertrand (APUR) for sharing his views on urban issues, both technical and political, and for his support and friendship throughout his professional experience in the world of urban planning.

References

Barles, S. (1999). La ville délétère: Médecins et ingénieurs dans l'espace urbain, XVIIIe–XIXe siècle. Champ Vallon.

Barles, S., & Thébault, E. (2018). Des réseaux aux écosystèmes: Mutation contemporaine des infrastructures urbaines de l'eau en France. *Tracés. Revue de Sciences humaines, 35*, Article 35. https://doi.org/10.4000/traces.8299

Baveye, P. C., Baveye, J., & Gowdy, J. (2016). Soil "ecosystem" services and natural capital: Critical appraisal of research on uncertain ground. *Frontiers in Environmental Science, 4*, 41. https://doi.org/10.3389/fenvs.2016.00041

Bispo, A., Guellier, C., Martin, E., Sapijanskas, J., Soubelet, H., & Chenu, C. (2016). Les sols Intégrer leur multifonctionnalité pour une gestion durable. Editions Quae. https://hal.archives-ouvertes.fr/hal-01531711

Blanchart, A., Séré, G., Cherel, J., Warot, G., Stas, M., Noël, C. J., Morel, J. L., & Schwartz, C. (2018). Towards an operational methodology to optimize ecosystem services provided by urban soils. *Landscape and Urban Planning, 176*, 1–9. https://doi.org/10.1016/j.landurbplan.2018.03.019

Clark, N. (2017). Politics of strata. *Theory, Culture & Society, 34*(2–3), Article 2–3. https://doi.org/10.1177/0263276416667538

Constant II, E. W. (1987). The social locus of technological practice: Community, system, or organization ? In W. E. Bijker, T. P. Hugues & T. J. Pinch, *The social construction of technological systems: New directions in the sociology and history of technology* (pp. 223–242). The MIT press.

Denis, J., & Pontille, D. (2020). Maintenance et attention à la fragilité. *SociologieS*. http://journals.openedition.org/sociologies/13936

Denis, J., & Pontille, D. (2022). *Le soin des choses: Politiques de la maintenance*. La Découverte.

Dumont, E., Branchu, P., & Sage, J. (2020). Faisabilité de l'infiltration des eaux pluviales en zone d'aléa dissolution du gypse—Cas de l'Ecoquartier de la Gare à Pantin (CEREMA). *Journées Nationales de Géotechnique et de Géologie de l'Ingénieur*.

Emelianoff, C. (2007). La ville durable: L'hypothèse d'un tournant urbanistique en Europe. *L'information Géographique, 71*(3), 48–65. https://doi.org/10.3917/lig.713.0048

European Commission. (2023). *Proposal for a directive of the European parliament and of the council on soil monitoring and resilience (Soil Monitoring Law)*. https://eur-lex.europa.eu/legal-content/EN/TXT/?uri=CELEX%3A52023PC0416&amp%3Bqid=1706624227744

FAO. (2015). Soil functions: Soils deliver ecosystem sevices that enable life on Earth—Infographics. *FAO*. https://www.fao.org/documents/card/fr?details=0815e457-c6a4-47e9-ab6c-f23224279834

Fournil, J., Kon Kam King, J., Granjou, C., & Cécillon, L. (2018). Le sol: Enquête sur les mécanismes de (non) émergence d'un problème public environnemental. *VertigO : la revue électronique en sciences de l'environnement, 18*(2), Article 2. https://journals.openedition.org/vertigo/20433

Knorr-Cetina, K. (2007). Culture in global knowledge societies: Knowledge cultures and epistemic cultures. *Interdisciplinary Science Reviews, 32*(4), 361–375. https://doi.org/10.1179/030801807X163571

Lin, B., Johnson-Chappell, M. J., Vandermeer, J., Smith, G., Quintero, E., Kerr, R., Griffith, D., Ketcham, S., Latta, S., Mcmichael, P., Mcguire, K., Nigh, R., Rocheleau, D., Soluri, J., & Perfecto, I. (2011). Effects of industrial agriculture on climate change and the mitigation potential of small-scale agro-ecological farms. *CAB Reviews: Perspectives in Agriculture, Veterinary Science, Nutrition, and Natural Resources, 6*, 1–18. https://doi.org/10.1079/PAVSNNR20116020

Mairie de Paris, Direction de l'Urbanisme. (2018, June 30). Flash Info. Guide d'application des dispositions environnementales du PLU [Personal communication].

Mairie de Paris. (2019, September 24). *Les eaux pluviales:Parispluie*. https://www.paris.fr/pages/le-plan-parispluie-5618

Menta, C., Conti, F. D., Pinto, S., Leoni, A., & Lozano-Fondón, C. (2014). Monitoring soil restoration in an open-pit mine in Northern Italy. *Applied Soil Ecology, 83*, 22–29. https://doi.org/10.1016/j.apsoil.2013.07.013

Meulemans, G. (2020). Urban pedogeneses: The making of city soils from hard surfacing to the urban soil sciences. *Environmental Humanities, 12*(1), Article 1. https://doi.org/10.1215/22011919-8142330

Nerland, M., & Jensen, K. (2012). Epistemic practices and object relations in professional work. *Journal of Education and Work, 25*(1), 101–120. https://doi.org/10.1080/13639080.2012.644909

Nezeys, A., Reboul, S., Saison, O., & Baillet, M. (2016, June 28). Le Plan pluie à Paris: La nécessaire dimension environnementale. *Novatech 2016—9ème Conférence internationale sur les techniques et stratégies pour la gestion durable de l'Eau dans la Ville*. https://hal.science/hal-03322076

Reid, W., Mooney, H., Cropper, A., Capistrano, D., Carpenter, S., & Chopra, K. (2005). Millennium ecosystem assessment. *Ecosystems and human well-being: Synthesis*.

Robertson, M. (2012). Measurement and alienation: Making a world of ecosystem services. *Transactions of the Institute of British Geographers, 37*(3), Article 3.

Zitouni, B. (2013). Asphalter ou désasphalter Bruxelles. Comment l'hygiénisme du passé rencontre l'écologie politique d'aujourd'hui. *Revue de l'Institut de Sociologie, 83*, Article 83.

CHAPTER 8

Geothermal Politics. Harvesting Deep Heat, interacting with the Bedrock

Alain Nadaï, Julien Merlin, and Olivier Labussière

List of Abbreviations

ADEME	French Energy Agency
AEC	Atomic Energy Commission
ANDRA	French National Agency for the Management of Radioactive Waste
BRGM	Bureau de Recherches Géologique et Minière (French Geological Service)
CEA	French Atomic Energy Commission

A. Nadaï (✉)
CIRED-CNRS, Cedex, France
e-mail: alain.nadai@cnrs.fr

J. Merlin · O. Labussière
Pacte Social Sciences Research Center, University Grenoble Alpes, CNRS, Sciences Po Grenoble, Grenoble, France
e-mail: julien.merlin@engie.com

O. Labussière
e-mail: olivier.labussiere@cnrs.fr

© The Author(s), under exclusive license to Springer Nature Switzerland AG 2025

O. Labussière et al. (eds.), *Back to the Ground*,
https://doi.org/10.1007/978-3-031-88888-5_8

EGS	Enhanced Geothermal System
HDR	Hot dry rock
HTG	High-temperature geothermal energy
IPGP	Paris Institute of Earth Physics
NATO	North Atlantic Treaty Organization

8.1 Introduction

This chapter looks at the history of high-temperature geothermal energy (HTG), a pioneering case of geoengineering that emerged in response to the oil crisis of the 1970s. During that decade, the crisis highlighted the danger of our reliance on fossil fuels. The oil shocks triggered a wide range of policy responses, including a surge in interest in renewable energies. In the race to find alternatives to fossil fuels, smaller technologies such as solar photovoltaics and geothermal energy were elevated to the status of global promise (e.g., CEC, 1974). Among these was HTG, an innovative form of geoengineering that aimed to tap into the Earth's crust—the granite basement or bedrock—to harness its intense heat.

At the time, the global geothermal potential was still poorly understood, but several geothermal anomalies, corresponding to basement outcrops, suggested that high temperatures could be accessed at the bedrock level, potentially enabling the production of thermal electricity on the surface. The vision that emerged sought to transcend surface anomalies by accessing the bedrock wherever possible, transforming deep geothermal potential into a national or international solution to the energy crisis. Technically, the approach involved fracturing the bedrock to create passages, injecting fluids, and establishing flow circulation to recover the deep rock's high temperatures at the surface.

This chapter traces the development of this vision, along with the associated scientific and technological concepts, as well as engineering practices (e.g., oil drilling, hydraulic fracturing, and nuclear waste burial) and scientific disciplines, that were supposed to enable the creation of these geothermal "loops." The analysis covers the concept of hot dry rock [HDR], which originated in Los Alamos, New Mexico, USA, in the 1970s and 1980s. This concept later evolved into the Enhanced Geothermal System [EGS] in Europe in the 1980s.

We follow the ups and downs of this technology, its international evolution, and transformations in response to the geological responses of the successive sites (caldera, horst) to which it was applied. From HDR to EGS, the (in)consistencies between the scientific vision of the deep rock and the practical experiences at the sites prompted disciplinary shifts (solid mechanics, geology, hydrogeology) and further explorations into the complexity of underground circulations, thereby opening up unprecedented geological knowledge.

We adopt a political geology approach to examine attempts at establishing flow circulations in bedrock. These attempts are viewed as ways for humans to connect with terrestrial intensities that have their own dynamics (Clark, 2011, 2013; Neyrat, 2018). As suggested by this literature, such dynamics may challenge our knowledge and control over them and the aspiration to create new productive "strata" in relation to them (Bobbette & Donovan, 2019; Clark, 2016; Kinchy et al., 2018; Labussière, 2021; Labussière et al., 2024; Yusoff, 2017).

We acknowledge that such strata include the formation of geo-socio-technical loops but are not limited to these, as they establish broader and often unknown connections with terrestrial intensities (Clark, 2011). From this point of view, geothermal loops are not continuous and closed systems, as envisioned by engineers. This analysis looks at them as articulations between heterogeneous and partly unknown terrestrial dynamics. We focus on the alignment(s) and mismatches, and their underlying reasons, between the scientific visions steering the experiments in HTG and the material responses from the sites and the underground. These are not indicators of success or failure but of human/terrestrial interactions.

The bedrock presents a unique case in the "political geology" literature, which has explored the emergence of new productive "strata" related to fossil capitalism and specific, usually prospected, sedimentary layers like coal, gas, or oil. Strictly speaking, the bedrock differs as it is defined by its contact with the overlying rocks. It is not confined in three dimensions, nor is it a "stratum" or a container to be emptied. It is a fissile and partly unknown material with its own responses (geometry, scale, temporality, intensity) and is challenging to govern (Dodds, 2020; Richardson & Weszkalnys, 2014). Loops that attempt to open, expand, and establish flow circulations for heat extraction are both trying to animate/activate the bedrock and to extract passive matter from it, thus only partially relating to extractive capitalism (Yusoff, 2013). The bedrock is not even a genuine geological category. Initially approached by HGT engineers through mechanical engineering, it was categorized as a homogeneous, passive, and inanimate heat reservoir and exchanger, and a domain of constraints. Within this vision, making the bedrock a component of the loop called for activating these constraints by stimulating fractures/circulation along its lines of tension (HDR).

The chapter tracks these loops being established in various parts of the world, including Los Alamos (USA) and Soultz-sous-Forêts (France). We explore how the vision of the bedrock evolved with the material responses it triggered at each site. While the EGS technology developed in Soultz-Sous-Forêts in the 1980s was directly linked to the HDR technology developed in Los Alamos in the 1970s and 1980s, the geological differences and material responses of these sites led to a shift in how the bedrock was perceived. As bedrock heterogeneities and openness to other underground circulations became evident, the functionality of the geo-socio-technical loops became challenging to understand, and the global geothermal promise seemed increasingly premised on its capacity to integrate with unknown underground dynamics. The chapter concludes by discussing the evolving interplay between loops and strata, drawing insights into our relationship with terrestrial heat and dynamics.

8.2 Los Alamos, Cradle of the Geological Concept of Hot Dry Rock

Hot-dry-rock experiments, first developed in the USA in the 1970s and 1980s, opened up to a new geological environment—the deep rock or basement—with the potential to generate high-temperature dry steam and electricity on a global scale.

Engineering Geothermal Reservoirs

Since the mid-1970s, geothermal projects have aimed to harness underground heat to generate electricity. However, successful projects were rare and limited to specific geological contexts. The limited potential—only about a dozen sites worldwide—prompted the search for an alternative: to tap the Earth's internal heat at greater depth (Smith, 1975).

This shift in vision involved creating reservoirs in the deep rock, managing flow passages to inject and circulate fluid, and recovering deep heat at the surface.

The development of these new loops and strata required experimenting at new sites with new knowledge (deep rock) and know-how, partly imported from other technical fields (nuclear, oil). It also involved building new networks and communities of players.

From Terrestrial Gradient to "Dry Steam Reservoirs" (Geothermal Anomalies) (Nineteenth–mid-twentieth century)

Peter J. Smith's (1975, 1983) account of the history of man's attempts to harness the natural heat of the Earth's crust bears witness to the long-standing challenges involved in doing so, especially when tapping deep heat and very high temperatures to generate electricity. Before the nineteenth century, mines were the primary access to the Earth's subsoil heat. While they indicated a hot terrestrial core, its nature and origin remained hypothetical (chemical reaction, Earth's formation). In the early twentieth century, the idea emerged of drilling a deep well to explore the Earth's thermal gradient, but it wasn't until the 1920s, when the oil industry designed the first drilling tools, that such an idea started to gain traction. However, rock is a poor heat conductor, and drilling a borehole did not provide a sufficient exchange surface to harness heat effectively. The use of explosives was considered, but the borehole diameters were too small to insert enough explosives to create adequate heat circulation. Creating circulation in a rocky environment seemed untenable.

A few alternatives involved exploiting the variations and singularities of the Earth's crust.[1] One was to consider wet rather than dry environments and install a heat exchanger in a floating cavity created in the middle of natural brine domes by dissolving salt with water. This idea drew inspiration from storing practices (hydrocarbons, compressed air, in the 1970s) in saline environments, which are better heat conductors. The second possibility was to use shallow natural reservoirs of dry heat (high temperatures), corresponding to geological anomalies. These were present in "rare" contexts mentioned by Smith, such as Italy (the Larderello site in Tuscany, operational since 1904) and the USA (Geysers field, California) (Smith, 1975). These "dry steam reservoirs" were hot enough to produce "dry" steam suitable for electricity generation in geothermal power plants. Although a few economically viable power generation sites, such as Larderello, existed in the 1970s, these natural reservoirs were few,

[1] Smith, M. C. (1979). *Future of hot dry rock geothermal energy systems* (No. LA-UR-79-683; CONF-790615-12). Los Alamos Scientific Lab., NM (USA).

and the actual potential of these anomalies was still unknown at that time (Smith, 1975: 10). The expansion of this technique thus depended on the ability of underground engineers to extract heat from the rock at greater depths, where it was ubiquitous.

Man-made Geothermal Reservoirs (1959–1970)

In the 1970s, engineers pursued this idea and started developing techniques to reach the Earth's bedrock and its high temperatures to *create* genuine "geothermal reservoirs": the "man-made reservoirs."

- Penetrating the Granite Bedrock and the Emergence of Hot Dry Rock (HDR)

As early as 1959, an engineer (R. Carlson) had suggested using nuclear explosives to increase the permeability of wells. Being smaller and more powerful, nuclear explosives could easily be introduced in boreholes. The fragmentation of the rock would produce a rubble chimney around the original cavity and establish communication with existing underground water tables, allowing access to underground hot fluids (the Plowshare concept, Fig. 8.1). However, the overall feasibility of the project faced skepticism regarding the effective flow of fluids that could be achieved, the low profitability of wet steam regarding energy production, and, above all, the risk of contaminating underground water tables with nuclear debris (Kennedy study, 1964).

Looking for dry cavities—that is, rocks without aquifers—to avoid the risk of nuclear contamination of existing water tables emerged as the solution, and the concept of "hot dry rock" (HDR) was born. It opened the door to a wide range of proposals.

In 1970, two engineers proposed adapting Carlson's idea to a dry environment by introducing surface water at the bottom of the cavity through a borehole (Fig. 8.1). At great depth, the high temperatures would convert the water into dry steam and pressurize it back to the surface. The valorization of dry steam in a power plant would offer a clear economic advantage. However, a detailed study pointed to the risk of radioactive contamination over time, due to the progressive degradation and dissemination of radioactive glass formations accumulated on the walls of the cavity at the time of the nuclear explosion.

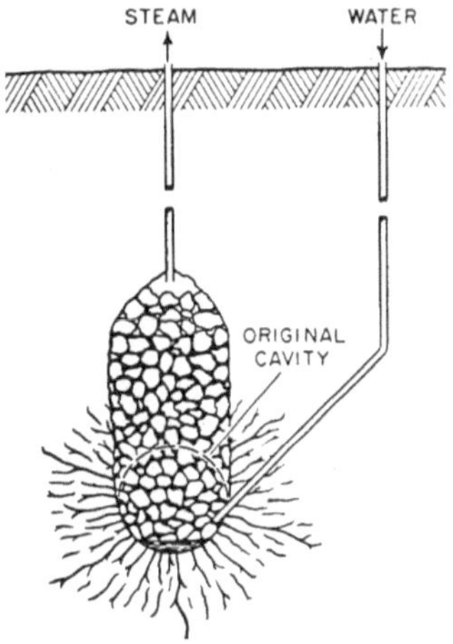

Fig. 8.1 Rubble-filled chimney produced by caving the roof into a cavity created by a nuclear explosive

- Melting the bedrock through nuclear penetration ("Subterrene")

At the same time, in 1970, an informal group of researchers at Los Alamos proposed another concept to penetrate the bedrock. Still relying on nuclear energy, the idea was to propel and heat a penetrator that would melt the rock: the Subterrene or "nuclear rock melting penetrator" (Robinson et al., 1971) (Fig. 8.2). According to its designers, this type of device could reach and drill geological environments difficult to access with conventional drilling techniques and was expected to become increasingly effective in extremely hot, great-depth environments, such as magma. Although it was never implemented, the concept challenged the idea that the granitic basement was an unattainable geological frontier and marked the emergence of HDR.

- Redefining the Resource and Designing Reservoirs into the Hot Dry Rock

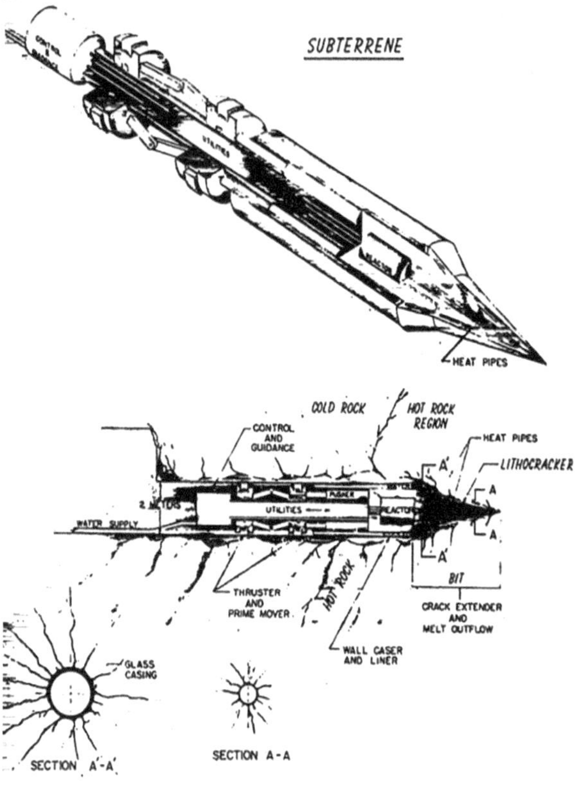

Fig. 8.2 The subterrene or "nuclear rock melting penetrator"

In the same year, the notion and harnessing of geothermal resources were associated with the existence of geothermal fluids.[2] Whether exploiting aquifers or dry steam, the presence of underground fluids was a precondition for harnessing geothermal energy and with it, the presence of a natural geological reservoir. HDR, although still only a concept and debated, deviated from the conventional vision of geothermal resources as a combination of fluid and reservoir, targeting the Earth's heat residing

[2] Robinson, E. S., Potter, R. M., McInteer, B. B., Rowley, J. C., Armstrong, D. E., & Mills, R. L. (1971). *PRELIMINARY STUDY OF THE NUCLEAR SUBTERRENE* (No. LA-4547). Los Alamos National Lab. (LANL), Los Alamos, NM (United States).

in the rock itself. In this new perspective, "nature" provided the heat, and human action the rest: the geological reservoir, the fluid (heat vector), and the sociotechnical assemblage enabling this fluid to circulate in a loop and the heat to be recovered at the surface.

The distinction between the fluid and the resource was based on three assumptions. First, it was possible to design impermeable reservoirs in the bedrock, hot enough to produce (dry) steam. Second, these reservoirs would be dry because no natural fluids were present in the bedrock. Third, it was feasible to create a flow circulation in these reservoirs, between the underground and the surface, so that liquid injected into these reservoirs could be recovered as (dry) steam at the surface.

As we observed, the resistance of the geological environment was a key factor in prompting engineers to consider dry environments. The resistance of the bedrock called for using (nuclear) energies, which made the circulation of natural fluids a source of contamination and overflows, thus complicating efforts to secure a closed loop and develop a productive *stratum*. However, designing hot and dry reservoirs requires much more sophisticated engineering than merely extracting geothermal fluids. Although several techniques were considered, the uncertainties associated with some of them gradually led to the exploration of techniques other than nuclear power, such as hydraulic fracturing.

Closing the Loop: Enrolling Sites, Containing Social Interactions, and Looping with the Bedrock (1970–mid-1980s)

Choosing Hydraulic Fracturing

In addition to early ideas of nuclear explosion (Plowshare concept) or propulsion (Subterrene), other techniques such as chemical leaching, chemical explosion, and hydraulic fracturing were also considered potentially effective, even though this effectiveness remained hypothetical (Fig. 8.3).

Several technologies were under consideration, and engineers were planning for widespread deployment beyond the confined and military site of Los Alamos. In so doing, they connected technological choice with the economic, environmental, and social conditions that would enable HDR to be deployed on a large scale.[3] The use of nuclear explosions was

[3] (Source: Smith, 1979, 1983).

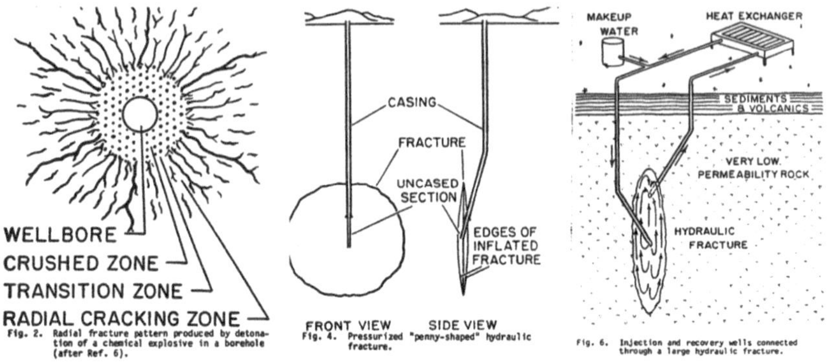

Fig. 8.3 Chemical (left) and hydraulic Fracturing (Center, Right) (Smith, 1979, 1983)

quickly dismissed, as engineers did not see the feasibility of deploying this technology outside of Los Alamos (Smith, 1979). From the beginning, the HDR system was influenced outset by strong anti-nuclear protests in the 1970s.

Conversely, hydraulic fracturing seemed to offer several advantages. Technically, it was easier to implement, as it was already part of oil and gas drilling protocols. It involved injecting high-pressure fluid to fracture the rock and access the resource. Adding various materials to the injected water also allowed keeping the fractures open during operation, which was key for circulating geothermal fluid. Hydraulic fracturing was also regarded as more natural and having greater "apparent environmental acceptability" since water was already present in nature (Smith, 1975). Despite a risk of induced seismicity, which was observed and anticipated by Los Alamos researchers (Smith, 1983: 14), it was seen as having low environmental impacts at the time.

These advantages, however, did not guarantee its feasibility for HDR. In their efforts, Los Alamos engineers encountered significant resistance from experts and scientists, who deemed HDR impossible (Smith, 1983: 14). To meet this challenge, the engineers had to enroll a full-scale geological site and establish a real-world field laboratory.

Looping the HDR and Enrolling a Site

- Hot Dry Rock Through Solid Mechanics

Engineers' conviction that hydraulic fracturing may allow them to loop a flow passage between the surface and the hot dry rock was based on solid mechanics. According to this, the granitic bedrock was a homogeneous, dry, and impermeable geological environment subject to directional compressive stresses inherited from the area's tectonic history.

Within this framework, pumping water into a wellbore would increase the pressure and build up tension, which would split the wall of the hole and produce a crack in the direction of the least stress and pressure. Further technical considerations led engineers to expect that the crack would be "vertical, planar, and normal to that stress axis" and "approximately circular in outline" (penny-shaped) (Fig. 8.3, center and right). In hydraulic fracturing, this process was generally facilitated by inserting "packers"—a pair of circular removable elastomeric seals—into the well.

This was the vision that Los Alamos engineers, starting with mathematical studies in the early 1970s and setting up an informal, partly volunteer "Los Alamos Geothermal Energy Group," decided to test in a caldera and former nuclear test site a few kilometers west of Los Alamos.

- Valles Caldera as a Field lab

Valles Caldera was chosen because of its suspected near-surface magma (hot springs), the absence of recent seismic episodes or faults around it, monitoring by the Geological Survey, and its public status. In 1972, four shallow boreholes (30 meters) were drilled to compare temperature gradients, which confirmed higher temperatures outside the western rim of the caldera. It was decided to drill a deep hole—the "Granite Test Hole No. 1" (GT1) at Barley Canyon—there to investigate the crystalline basement. GT1 reached a depth of 785 m, penetrating 143 meters into the Precambrian bedrock and encountering a temperature of 100.4 °C. A series of experiments in hydraulic fracturing were conducted with the support of the Atomic Energy Commission (AEC). The resulting fractures were vertical and directional, and the loss in fluids was very limited, validating the engineers' vision and indicating that the Precambrian was suited for establishing a flow passage. To test a larger circulation, another deeper hole (GT-2, 2928 m, 197 °C) was drilled two years later a few kilometers south of GT-1, in Fenton Hill, an area with favorable access and working conditions (road access, no timber/burned land, electrical and telephone line, public status). The conditions to contain a pressurized-water loop were shown to be prevalent in this deeper and somewhat distant environment. The stratum of the bedrock was confirmed to support the geothermal loop.

The results allowed the group to gain more substantial institutional and financial support (US Energy R&D Administration, US Department of Energy) and to drill another deep hole (the Energy Extraction Hole 1 or EE-I) connected to GT2, which included a pressurized-water heat-extraction loop that operated for several months. The fracture was then enlarged, and the system operated for a longer period, setting the stage for a larger and deeper experiment in 1983. With the geothermal loop happening again and persisting, the concept of a homogenous and impermeable bedrock was confirmed, merging the bedrock as a stratum with the geothermal loop.

Based on these successes, the US Department of Energy decided in 1978 to expand the Los Alamos project into a nationwide HDR program by 1983 (Smith, 1983).

Disseminating HDR: Site Identification and the Deep-Rock Technopolitics

While developing their field lab, the Los Alamos engineers continued initiatives to convince a broader expert audience of the value of their experiments and secure institutional support. Soon, under the impetus of the US, HDR became an international topic, and its implementation extended beyond the confined US military-scientific sites.

On a technical level, HDR was backed by an expanding engineering community, drawing on multiple fields of knowledge (e.g., solid mechanics) and engineering disciplines (geothermal, nuclear, oil, mining). As the oil shocks of 1973 and 1979 highlighted the fragility of dependence on fossil resources, HDR embodied a promise of global transition. It evolved into a genuine technopolitics aimed at developing bedrock engineering (solid mechanics, nuclear experiments, oil drilling and hydraulic fracturing practices, etc.). Few if any private players were involved in it; instead, states and their public geological services, NATO, and the Commission of the European Communities support were at the core of it.

In 1973, US representatives steered a process within a NATO committee to identify geothermal pilot sites around the world, aiming to establish international cooperation on HDR among these sites. These exchanges led to cooperation between certain states, particularly Italy and the USA, to compare man-made and natural reservoirs.

In the European Community, the "Community Energy Research and Development Programme" was set up in 1975 to assess Europe's geothermal potential and help member states establish national geothermal programs and data inventory. The first two versions of this program were partly dedicated to HDR and identifying sites with geological similarities to Los Alamos. About 20 pilot sites were identified. HDR experiments were financed within this program at Rosemanowes (Cornwall) and Falkenberg and Urach (Germany). The third version of the program supported the Franco-German experimental project at Soultz-sous-Forêts (France). In France, as the initial test sites were also nuclear waste storage sites, HDR and hydraulic fracturing became intertwined with nuclear issues and their opposition.

8.3 Soultz-sous-Forêts, a Geological Concept Challenged by the Bedrock

In search of granite outcrops (1975–1989): the emergence of deep geothermal energy in France.

Le Mayet-de-Montagne: the escaping flows of an outcropping granite

In 1973, France was absent from NATO committee discussions on HDR and site identification, like most European countries, which were focused on discussions about Dogger geothermal energy.

By 1977, the situation had changed. At another NATO HDR meeting in Los Alamos, BRGM announced that European funding had enabled the start of laboratory experiments with bedrock fracturing in France. These experiments involved chemical leaching, which was used in the mining industry. One year later, the European Geothermal R&D program allowed researchers from the Paris Institute of Earth Physics, including François Cornet, to launch real-site experiments in hydraulic fracturing for HDR in Le Mayet-de-Montagne. These experiments engaged BRGM teams interested in the conductivity of hydraulic fractures[4] and connected with communities of researchers and engineers interested in bedrock, fluid

[4] Researchers B. Feuga, M.L. Noyer, and O. Thiery.

flow modeling, and heat transfer in fractured environments.[5] These were critical research areas for various technical fields such as nuclear waste storage, coal gasification, enhanced oil recovery, and oil shale recovery.[6]

Le Mayet-de-Montagne was selected because of the presence of outcropping granite on the site. The aim was not to produce heat or electricity but to use sensors to record seismic vibrations (geophones[7]), understand how hydraulic fracturing affected the bedrock, and model how fractured faults worked. The experiments were partially successful. Although the granite was effectively fractured, the faults' conductivity remained a mystery. Various substances were injected into the fractured reservoir (sand, fluid), but the engineers were unable to "pressurize" the subsoil[8] nor recover at the surface the substances they had injected.

The results challenged the HDR concept and technology developed in Valles Caldera, breaking with the idea that it was possible to artificially create fractured reservoirs in low-permeability environments. They also led to the realization that the "granitic basement" was a homogeneous environment: Results obtained at one site proved difficult to extrapolate to another site. More precisely, François Cornet's work pointed to the existence of pre-existing cracks in the granitic basement,[9] which shifted the problem from knowing whether it was possible to create an artificial reservoir in this basement to interpreting the channeling between pre-existing and newly created cracks caused by fracturing.

Cornet's work initiated a change in approach that would later be applied to other experimental sites, including Soultz-sous-Forêts. The idea was no longer to create reservoirs in the bedrock but to improve flow circulations in it by stimulating and "opening [of] pre-existing fractures" (*ibid.*). Facing the resistance of a new site, the HDR concept was

[5] Centre d'Informatique Geologique de l'École des Mines de Paris Fontainebleau, ingénieurs Ledoux et Hosanski.

[6] Leblanc, P. (1986). Structure et hydrodynamique des milieux fissurés aquifères. BRGM report 86 SGN 102 EAU.

[7] A geophone is a device that converts ground movement (velocity) into voltage, which can be recorded. Data interpretation, such as when the ground is "stimulated" (e.g., fractured), enables modeling of the underground and its various components and geophysical behavior.

[8] Interview conducted by the authors with a BRGM expert (2021).

[9] Cornet, F. H., & Descroches, J. (1990). The problem of channeling in hot dry rock reservoirs. *Hot Dry Rock Geothermal Energy*, 398–421.

redefined, as was the bedrock as a geological environment, which was no longer seen as homogeneous but heterogeneous. Thus, to be completed, the geothermal loop should not close in on itself but interact and negotiate with a complex and potentially open network of pre-existing faults: a *stratum*.

Nuclear Issues Politicizing the Granite Outcropping

In France, the advent of geothermal energy in the 1980s coincided with the emergence of burying nuclear waste from power plants or former uranium mines on the policy agenda, and the subsequent interest in the subsoil as a storage site. Whether clay, salt, shale, or granite, the bedrock was central to the interests and initiatives of groups of engineers at the French Geological Service (BRGM), the French National Agency for the Management of Radioactive Waste (ANDRA), the Atomic Energy Commission (CEA) and its subsidiaries involved in nuclear fuel management (CEA-Industry, Cogema), and academic departments (Paris Institute of Earth Physics [IPGP], laboratories of the Paris School of Mines).

About 20 kilometers from Le Mayet-de-Montagne, the Mine des Bois Noirs, operated by Cogema, was exploiting the granite outcrops for their rich uranium content (Cuney, 1978). When the mine closed, CEA-Industrie planned to convert the mine into a nuclear waste storage site. The project fell short (Blanck, 2017), but local groups, including the "Collectif des Bois Noirs," began protesting against it in the late 1970s, coinciding with the Le Mayet-de-Montagne project's fracking experiments a few kilometers away. The experiments clashed with the local town council, and local residents mobilized against the storage project.[10] Thus, the bedrock became a central geological environment not only for public policies but also for new publics who demanded public debates on its uses, thus integrating it into regional planning issues. Deep geothermal explorations had to be carried out in an area and on a site with a different heritage.

[10] Interview conducted by the authors with a BRGM expert (2021).

Soultz-sous-Forêts and the Bedrock of the Oil Legacy

In 1984, an inventory campaign conducted by the French Energy Agency (ADEME) and the BRGM identified a thermal anomaly, indicative of a basement outcrop, at Soultz-sous-Forêts (Alsace, Eastern France). Situated in the Rhine Graben, a tertiary rift, the region boasted a rich history of subsoil exploration and exploitation, particularly for oil. Notably, the site's proximity to the Pechelbronn oil field, where oil shale had been mined since the Renaissance, added to its significance.

Similar to the Paris Dogger Basin, the petroleum legacy at Soultz made its subsoil a distinctive area that was already partially understood.[11] It provided BRGM engineers crucial insights into the sedimentary cover and the "top of the granite basement." By leveraging vibroseismic data provided by Total and information from previous drillings, BRGM engineers determined the depth of the Palaeozoic granite. They identified a significant compartment, or "horst," adjacent to the "graben," beneath which the granite bedrock was accessible at a shallower depth. The petroleum data also proved indispensable for drilling the initial exploration borehole GPK1, which reached a depth of 2000 meters and revealed a temperature of 140 °C.

Starting in 1988, numerous injection tests were conducted in GPK1, experimenting with various chemicals to trace the circulation of the fluid. Both artificial and natural tracers were used to gain a comprehensive understanding. Temperature measurements, flow assessments, and hydraulic tests were carried out in tandem (Schellschmidt & Schulz, 1991). These injection tests were pivotal in gauging the resistance of the geological environment.

At that stage, the primary objective of the engineers was not to establish a geothermal loop; no doublet was in place. Instead, their goal was to enhance understanding and characterize the basement. A notable feature of this period was the tightly knit technical community, deeply rooted in this specific site and geological context. Knowledge exchange was intricately linked to the environment and the wealth of insights gathered since the petroleum era, as well as the infrastructure inherited from that time. A geotechnical common ground was emerging, fostering collaboration among various stakeholders involved in subterranean activities.

[11] Dezayes, C., Gentier, S., & Genter, A. (2005). Deep Geothermal Energy in Western Europe: The Soultz Project: Final Report (BRGM/RP-54227-FR). BRGM.

It was precisely this embedded knowledge that facilitated the identification of the geothermal potential—a geological know-how considered in anthropology (e.g., Richardson & Weszkalnys, 2014).

In the early 1990s, an old oil well was reactivated, and an attempt was made to extend its depth, resulting in the creation of EPS1. The primary objective of this borehole was to conduct comprehensive hydrological tests at greater depths, particularly in preparation for a complementary borehole to GTK1 (referred to as GTK2), to establish a geothermal doublet. Despite encountering deviation issues that prevented it from reaching its intended depth, EPS1 yielded crucial data points and provided a nuanced understanding of the granitic basement. BRGM engineers Genter and Traineau (1992) translated these results into a three-type classification for the "granite environment," depending on the presence (or not) and the openness/activity (or not) of the fractures:

> from the viewpoint, of developing an HDR reservoir at the Soultz site, the granite can be classfied into three main types: a) 'standard granite' with a low natural fracture density, which corresponds to the standard model of the hot dry rock reservoir; b) fractured and hydrothermally altered zones related to paleohydrothermal systems, similar to self-sealed fracture reservoirs observed in natural geothermal systems; and c/ fractured and altered zones that are still active ans related to a present-day hydrothermal system.[12]

This description of the bedrock not only underscored the implications of François Cornet's results at Le Mayet-de-Montagne but also marked the onset of a paradigm shift in the characterization of the granite environment. Unlike the original Los Alamos–derived HDR concept, this environment was now acknowledged as naturally faulted and "wet." This pushed the boundaries of existing knowledge and prompted a re-evaluation of geothermal exploration strategies.

The Search for a Geothermal Loop in an Open Environment

The Soultz project, spanning from 1990 to 2000, saw BRGM engineers deepen the GPK1 well to 3600 meters in 1990 to conduct "injection tests" and refine the characterization of the geological environment.

12 (Source: Genter & Traineau, 1992).

Advances in micro-seismic technology throughout the 1990s enabled the development of increasingly sophisticated models capable of simulating fluid flows within complex geological fracture networks.

In Soultz, a series of rigorous tests challenged the assumptions underlying HDR technology. During the fluid injections, geophones were strategically placed in nearby abandoned oil wells to capture seismic data and interpret the porosity of the geological formations. These tests revealed unexpected complexities and inconsistencies. The geological environment proved to be more dynamic than previously thought, and the BRGM engineers had to fundamentally reconsider their initial conceptions. They began to shift their focus toward concepts such as "HWR" (Hot Water Rock) and "HFR" (Hot Fractured Rock) to reflect their evolving understanding of the geological conditions conducive to sustainable geothermal energy extraction. This new perception was further confirmed by subsequent drilling activities.

In the early 1990s, an experimental phase focused on the technical capabilities required for drilling at depths of between 2000 and 3000 meters into the granite rock. To facilitate this phase, drilling expertise from the USA and ENEL was enlisted. A new well, GPK2, was drilled close to EPS1. GPK2 was slated to serve as the primary well for a geothermal doublet. The goal was to establish or enhance a connection between the two wells, GPK1 and GPK2. From 1995 until 1997, multiple hydraulic fracturing tests were conducted in both wells to enhance the geological environment's permeability, and the first circulation tests between the two wells occurred. Yet, the loop did not materialize as expected: The test revealed uncertainty about the very nature of the underground.

Monitoring Flows, the Bedrock as Hydrogeological System (EGS)

Circulation tests used tracers—substances injected into one of the wells and partly recaptured by the engineers at the other well. The speed and rate of recovery of the tracers at the second well were used to assess the geothermal loop.

A four-month-long circulation test revealed that only 30% of the fluid injected into the system was recovered, suggesting that 70% of these fluids correspond to geothermal fluid already present in the geological environment. Thus, the Soultz system was an open geothermal system, connected and articulated with terrestrial loops and fluids.

Between 1998 and 2005, two additional wells were constructed: GPK3 and GPK4. The primary objective was to reach greater depths and test new stimulation methods, including chemical stimulation, to enhance the fluid flow from injection to recovery wells. These experiments culminated in what BRGM engineers termed "environmental enhancement" to refer to operations including fracture shearing and the dissolution of calcites within the environment. Importantly, the concept acknowledged the existence of geothermal rhythms and loops intrinsic to the environment itself and suggested composing with it. Such a compositionist approach stood in stark contrast to the notion of a "man-made reservoir," which presupposed complete control over both the reservoir and the flow circulation. To capture the change and pioneer this innovative approach, BRGM and the Soultz project introduced the concept of an Enhanced Geothermal System (EGS).[13]

Importantly, EGS challenged the underlying assumption of scalability that underpinned the shift to dry environments and the emergence of HDR technology. Composing with a heterogeneous and open geological environment was very different from looping with a dry and close geological reservoir. Composing acknowledged heterogeneity and uncertainty in the interaction with the bedrock, which complicated straightforward technological replication. Flagging a basement outcrop would no longer guarantee the feasibility of a geothermal loop because of the necessity to engage with the underground as a complex, heterogeneous entity, requiring collaboration across diverse technical communities and knowledge. Listening to the underground, to understand its geomechanical "behavior,"[14] became a central issue. The loop's feasibility depended on *negotiating with the stratum* to stabilize geothermal production and the economy (Table 8.1).

8.4 Geothermal Politics

At first glance, when considering geothermal politics, the relationship between the socio-material loop installed in each case study and the stratum—i.e., the interaction with the bedrock that results from attempts

[13] A large part of this history is based on interviews with BRGM engineers and on the report BRGM/RP-54227-FR written by C. Dezayes (2005).

[14] The notion of "behavior" is used by the engineers themselves working on the subsoil at Soultz-Sous-Forêt. See for example the source cited above.

Table 8.1 Exploration of the bedrock in Soultz-sous-Forêts (*Source* Authors)

Stages	*Dates*	*Geotechnological principles*	*Details*
Stage 1: Exploring the geological environment (GPK1)	1987–1991	HDR concept: "the "man-made" concept	The granite bedrock is perceived as watertight and homogeneous. It needs to be fractured to create a reservoir
Stage 2: Attempt at creating a geothermal loop (EPS1 / GPK1 / GPK2)	1991–1998	Re-conceptualization of the geological environment (hot fractured rocks & hot water rock concept)	–Experiments do not meet the HDR concept –Natural fractures do exist; fluid flow in faults is difficult to interpret –Assumes an environment already containing natural fluids
Stage 3: Toward an understanding of a complex geological environment (GPK1, 2, 3, 4)	1999–present	Re-conceptualization of the geotechnological principle. Innovation process that integrates the specificity of the underground: enhanced geothermal system	–Characterization of a naturally faulted geological environment –Concept of a reservoir in a homogeneous environment is jeopardized

at harnessing the Earth's inner heat—seems quite different.[15] In Los Alamos, the strata seem to be conflated with the loop. The flow circulation appears to somewhat encapsulate the interaction with the bedrock. Allegedly, entrant fluids are channeled in and recovered from the man-made reservoirs, to an extent that supports public investment and scaling up of the experiment. In Soultz-sous-Forêts, the flow circulation, although quantitatively balanced, eventually interacts with unknown material flows and exchanges, which seem to far exceed what engineers can trace. The stratum encompasses and extends well beyond the loop. While giving way to an economically feasible extraction of high heat,

[15] Source: authors.

the relationship with the bedrock remains somewhat transactional and open-ended, neither delimited nor circumscribed.

These case studies thus point to different articulations and interactions with terrestrial intensities: In the first, the interaction seems confined, and the bedrock is watertight and homogeneous; in the second, the interaction is not confined, and the bedrock is wet, faulted, and open-ended. This difference also reflects differences in scientific and engineering communities and knowledge, with a shift from solid mechanics to hydrogeology.

These loops are installed in different geological environments, chosen with careful consideration, and seem decisive for the fate of the stratum. The proximity of the bedrock, the presence of surfacing heat anomalies, and the accumulation of knowledge about the underground (geological survey, nuclear testing, and oil exploitation) are all crucial for locating the experiments in Valles Caldera, in one case, and in the Rhine Graben, in the other case. In Los Alamos, the absence of recent seismic episodes or faults is deemed essential through the prism of solid mechanics and influences the choice for the periphery of the caldera. In Soultz-sous-Forêts, the proximity of the bedrock drives the selection of a horst within the Rhine Graben. The faulted structure of this underground likely contributes to the shift in scientific paradigm from considering the bedrock as a matter in tension (solid mechanics) to representing it as a composite environment (hydro-geothermal system). Thus, there is likely a role played by the underground in steering human exploration and human–Earth interactions.

The adjustment of access and knowledge also reflects a relationship between the inner and outer Earth at work. A circulation between geothermal and geological politics seems to play out in the practices of geothermal project developments. Discussing the connection between geologic and geothermal politics, Clark suggests a recent shift in geological politics from "issues of territorial divisions of the earth surface to issues concerning the strata that compose the deep temporal earth" (Clark, 2013). With this shift comes the question of what is within our reach or under our responsibility and what extends beyond the political per se.

Clark highlights a shift in our responsibility. As the concept of the Anthropocene underlines our involvement in changing terrestrial dynamics, we have moved from *determinism*—where Earth dynamics dictated the possibility of our life on Earth to *possibilism*, where we also help shape these conditions. With this shift, the issue is raised of how the consequences of our actions on terrestrial dynamics are readable at the Earth's surface, allowing us to account for them.

The bedrock is a border (line) geological category in several ways, bridging surface geothermal objects and deep Earth dynamics. First, as shown by the open-ended fluid circulations and stratum in Soultz-sous-Forêts, trying to reach the bedrock involves interacting with inner terrestrial dynamics that remain largely unknown. These dynamics may stay beyond our reach even as we interact with them, mediated by successive flows that are unfamiliar to us.

Second, concerning deep heat, the heat that we harvest by interacting with the bedrock comes from magma pockets in the mantle, which are not well understood or located. Even if geothermal energy is classified as renewable energy in our institutions, deep Earth heat is not renewable per se; it is presumed to be present in such quantities that our extraction does not affect deep Earth dynamics. In this respect, indirect interaction with magma, as mediated by the geological environments or geothermal dynamics in our case studies, is very different than what direct interaction with magma might require in terms of precautions or prospects (Clark et al., 2018).

Ultimately, the case studies highlighted clearly display the articulations between surface politics of geothermal energy or other energies and the indirect and uncertain interactions with the geological underground. They show that the connection between our actions and the deeper terrestrial strata engaged by the loops we install is mediated by the underground itself. Thus, it is not within our reach to fully gauge and control the consequences of our interactions with terrestrial dynamics, even if we want to do so.

Competing Interests and Acknowledgements This work was carried out with the financial support of the CVT Ancre through the research program "Etude des conditions pour un développement soutenable de nouvelles technologies de l'énergie: la géothermie et le stockage de chaleur / froid dans le sous-sol," coordinated by O. Labussière and A. Nadaï.

Bibliography

Blanck, J. (2017). *Gouverner par le temps: La gestion des déchets radioactifs en France, entre changements organisationnels et construction de solutions techniques irréversibles (1950–2014).* (Doctoral dissertation). Institut d'études politiques de Paris-Sciences Po.

Bobbette, A., & Donovan, A. (2019). Political geology: An introduction. In A. Bobbette & A. Donovan (Eds.), *Political geology: Active stratigraphies and the making of life.* (pp. 1–34). Palgrave.

Clark, N. (2011). *Inhuman nature.* Sage.

Clark, N. (2013). Geoengineering and geologic politics. *Environment and Planning A, 45*(12), 2825–2832.

Clark, N. (2016). Politics of strata. *Theory, Culture & Society, 34*(2–3), 211–231.

Clark, N., Gormally, A., & H. Tuffen (2018). Speculative volcanology: Time, becoming, and violence in encounters with magma. *Environmental Humanities, 10*(1), 273–294.

Commission of the European Communities. (1974). *Energy for Europe: Research and development.* SEC (74)2592 final, Appendix II.

Cuney, M. (1978). Geologic environment, mineralogy and fluid inclusions of the Bois Noirs Limouzat uranium vein, Forez, France. *Economic Geology, 73*, 1576–1610.

Dodds, K. (2020). Cracking, forcing and covering up. In P. h. Billé. *Voluminous states: Sovereignity, materiality and the territorial imagination* (pp. 105–118), Duke University Press.

Genter, A., & Traineau, H. (1992, January). Hydrothermally altered and fractured granite as an HDR reservoir in the EPS-1 borehole, Alsace, France. In *Seventeenth workshop on geothermal reservoir engineering*. Stanford University.

Kinchy, A., Phadke, R., & Smith, J. (2018). Engaging the underground: An STS field in formation. *Engaging Science, Technology, and Society, 4*, 22–42.

Labussière, O. (2021). A coalbed methane 'stratum' for a low-carbon future: A critical inquiry from the Lorraine Basin (France). *Political Geography, Special Issue "Earth Politics", 85*, 102328

Labussière, O., Merlin, J., & Nadaï, A. (2024). Orchestrating deep geothermal energy policy. The exploitation of the dogger aquifer in the Paris Basin (1950–1990). In P. H. Hamman & N. Roudil (Eds.) *The future of cities and energies in Western Europe*. De Gruyter.

Neyrat, F. (2018). *The unconstructable earth.* Fordham Press.

Richardson, T., & Weszkalnys, G. (2014). Introduction: Resource materialities. *Anthropological Quarterly, 87*(1), 5–30.

Robinson, E. S., Potter, R. M., McInteer, B. B., Rowley, J. C., Armstrong, D. E., Mills, R. L., & Smith, M. C. (1971). A preliminary study of the nuclear Subterrene. *Los Alamos Scientific Laboratory Report, LA-4547*, 62.

Schellschmidt, R., & Schulz, R. (1991). Hydrogeothermic studies in the hot dry rock project at Soultz-sous-Forets. *Geothermal Energy in Europe: The Soultz Hot Dry Rock Project, 3*(1–4), 217–238.

Smith, M. C. (1979). The future of hot dry rock geothermal energy systems. *American Society of Mechanical Engineers, 79-PVP-35*, 12.

Smith, M. C. (1983). A history of hot dry rock geothermal energy systems. In: G. Heiken & F. Golf (Eds.), Geothermal energy from hot dry rock. *Journal of Volcanology and Geothermal Research, 15*, 1–20.

Smith, P. J. (1975). Towards universal geothermal power. *Nature, 257*(4), 10–11.

Yusoff, K. (2013). The geoengine: Geoengineering and the geopolitics of planetary modification. *Environment and Planning A, 45*(12), 2799–2808.

Yusoff, K. (2017). Geosocial strata. *Theory, Culture & Society, 34*(2–3), 105–127.

CHAPTER 9

Constructing Resources for the Transition: Geo-legal Approaches to Lithium-Bearing Brines in South America's Salt Flats

Marie Forget, Vincent Bos, and Chloé Nicolas-Artero

List of Abbreviations

ADI	Áreas de Desarrollo Indígena—Indigenous Development Areas
CCHEN	Chilean Nuclear Energy Commission—National Nuclear Energy Commission

M. Forget (✉)
Savoie Mont Blanc University, EDYTEM UMR5204, Le Bourget du Lac, France
e-mail: Marie.Forget@univ-smb.fr

V. Bos
CEA/DES/I-Tésé, Université Paris Saclay, Gif sur Yvette, France

C. Nicolas-Artero
Institut de Recherche pour le Développement UMR SENS, Montpellier, France

© The Author(s), under exclusive license to Springer Nature Switzerland AG 2025

O. Labussière et al. (eds.), *Back to the Ground*,
https://doi.org/10.1007/978-3-031-88888-5_9

CORFO	Corporación de Fomento de la Producción—Production Development Corporation
CSR	Corporate Social Responsibility
DGA	Direction General de Aguas—General Water Board
LIBs	Lithium-ion batteries
OCMAL	Observatorio de los Conflictos Mineros en America Latin—Mining Conflicts in Latin America Observatory
SMA	Secretaria del Medio Ambiente—Secretary of the Environment
SQM	Sociedad Química y Minera

9.1 Introduction

As it relies on exploiting new resources and developing infrastructures, the energy transition is fundamentally material (e.g., Bridge, 2009; Bridge et al., 2013; Labussière & Nadaï, 2020; Nadaï & Van der Horst, 2010). By focusing on extractive spaces related to this transition, we can understand the roles and strategies of key actors across different scales of the energy sector's reconfiguration. Furthermore, analyzing socio-ecological transitions through the prism of lithium sheds light on the de-territorialization processes at the scale of the territories and the involvement of surface and underground strata in global production processes. In the Andes, salt flats known as *salares* are closed drainage basins that resemble wetlands with a superficial salt crust, located in the arid, high-altitude Andean plateaus called the Altiplano, which features a vegetation formation called Puna. These deposits span the borders of three countries—the south of Bolivia and the north of Argentina and Chilecollectively referred to by the global industry as the "lithium triangle," a term that problematically focuses solely on lithium, thereby overshadowing other socio-economic and cultural potentials (Forget & Bos, 2022). This region occupies a central role in global production networks, serving as a productive node that hosted more than one-third of global production in 2023 and more than half of the world's brine reserves (USGS, 2020: 99).[1] In terms of exploitation, surface activities

[1] Chilean and Argentine production accounts for roughly 30% and 5%, respectively, of the 130,000 tons.

mobilize deep aquifers where brines[2] are pumped to the surface and evaporated in large pools in an extraction cycle lasting around 18 months. While mineral concentrations vary from one salt flat to another, the mechanisms of lithium accumulation are expected to be consistent (Houston et al., 2011).

This circulation of lithium-bearing brines reconfigures the organization of surface territories based on the development of underground resources, despite the under-studied and under-documented dynamics of these aquifers.[3] Traditional relations between inhabitants and their environment are reconfigured, endowing companies with new scientific and legal expertise.

Since the mid-2010s, global lithium flows have experienced acute growth driven by a boom in the production of lithium-ion batteries (LIBs) (Colchico, 2020). Electrification and the use of "smart systems" have placed LIBs at the center of sociotechnical solutions (idem; Olivetti et al., 2017; Sun et al., 2017). This extractive industry creates miningscapes resulting from the development of territories that serve extractive processes, most of which are orchestrated from outside the territory (Mendez et al., 2020). Lithium miningscapes are multiple, evolving in time and space, and are both visible and invisible (Fig. 9.1). The first condition for the construction of lithium miningscapes resides in the invisible composition of the underground: They are formed and informed by geology. The culturalist approach asserts that natural resources are not a given but social constructions, framed by their sociotechnical contexts (Bridge, 2009; Raffestin, 1980). Thus, lithium miningscapes associate nature (deposits) and culture (resource social construction). In nature, lithium is never found in its native state in metallic form. It can be dissolved in a liquid (as in seawater or geothermal water, for example) or be present in solid form within the crystal lattice of certain minerals. These landscapes replay the various geological strata laid down over the Earth's long history, aligning them with the short timeframe of resource exploitation, mostly dictated by the volatile prices of a global market. In fact, *salares* are an evaporation landscape created by the combined action of the dissolution of volcanic salts and evaporation

[2] In Argentina, the law defines brines as water resources containing more than 3.5% of dissolved salts (López Steinmetz & Bing Fong, 2019).

[3] Interview with Mr. Cervieto, hydrogeologist employed by the Atacameño Peoples Council, San Pedro, November 2019.

of high mineral-laden water over millions of years. Lithium exploitation attempts to reassemble the temporalities of capital by aligning productive tools with other geological (brines) and environmental (sun, water) temporalities.

This chapter examines the development of Chile's lithium industry from a legal geography perspective (Blomley, 1994; Braverman et al., 2014). This field offers an interdisciplinary approach that combines geography with law to "*question links between law, space, and power in order to understand how the production of space is associated with transformations of the political economy*" (Nicolas-Artero, 2020: 73). The aim is to explore the social production of natural resources, a multidimensional lithium-bearing space, and, more broadly, socio-ecological transitions through the prism of "*geo-legal actions, defined as the set of actions for reappropriating law and space that participate in the production of the latter*" (*idem*: 56). These actions are developed by three categories of actors: the State, companies, and local populations. Specifically, we will examine geo-legal tactics and strategies that differ based on the positions of actors within local power dynamics. The first includes "*the set of actions, with selective application, of diversion and of production of legal norms, carried out by actors, situated in a position of inferiority within local power relations,*

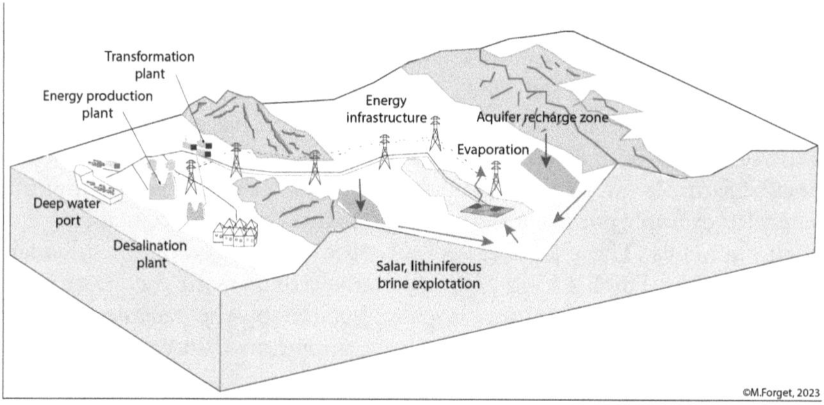

Fig. 9.1 The Salar de Atacama is a source of underground brines, formed by natural leaching from the Andes mountain range. Over time, the different minerals found below the salty surface crust have descended from the mountains and accumulated in growing concentrations

to appropriate space to resist the expansionof capitalism" (Nicolas-Artero, 2020: 59–60).

Several recent studies on mining analyze the role of law in the development or contestation of these industries. One approach reflects on the social construction of mining law and its spatial effects. Mining law can be analyzed as a colonial project, imbued with a liberal conception of property that favors the separation of land into two categories (surface and subsurface), thereby enabling exploitation through mining at the expense of private owners and indigenous peoples (Hoogeveen, 2015, Merveille, 2011, Forget et al., 2018). Extractive projects rely on an interweaving of actors and legal decisions that promote their rapid growth (Bos, 2017; Forget & Carrizo, 2016; Nicolas-Artero, 2020; Turton, 2015). Often, mining legislation and illegal practices for controlling mining resources contradict the recognition of local communities' territorial rights (Ulloa, 2020; Velez-Torres, 2014). Another approach mostly focuses on how protest movements against mining refer to the law, extending to the reflection on environmental justice. Social mobilizations suggest a regulatory framework for obligating the rehabilitation of closed mines and denounce the difficulties of its application (Toumbourou et al., 2020).

Our chapter expands discussions concerning the legal dimension of nature's social construction (Delaney, 2001; Marusek, 2021). We analyze its different conceptions from the perspectives of legislators, companies, and inhabitants by investigating their geo-legal positions in processes of re/de/constructing "lithium resources," "brines," and "water" that seek to favor or restrict the appropriation of space and secure control over it. We also contribute to discussions on how the law is mobilized in environmental conflicts (Bottaro & Sola Álvarez, 2016; Melé et al., 2020), showing how these conflicts inform the social construction of space and its rules.

The methodology used includes an analysis of legal texts and semi-structured interviews conducted with members of the Atacama "La Grande" communities, representatives of extractive companies, and local governments (November 2018). The creation of a corpus of legal texts and its hermeneutic analysis allows us to interpret and restore the meaning of law for those who produce it (Santoire et al., 2020). Finally, to update this data, a survey of Chilean and specialized (mining) press complements this material. However, this research does not take into account the latest agreements between SQM and Codelco at the time of publication in June 2024.

First, we will demonstrate how legal frameworks construct lithium resources to facilitate spatial organization, separating soil and subsoil to benefit the State. Next, we will delve into the geo-legal strategy used by mining companies, which differentiates between "brine" and "water" as distinct resources. Finally, we will show how local populations employ geo-legal tactics to reclaim territorial control and counter the expansionist strategies of both companies and the State.

9.2 Geo-legal Strategies for Organizing an Extractive Territory

Any resource is the unstable result of a geographically situated and historically dated social process that is inscribed in a sociotechnical system (e.g., Grieco & Salazar-Soler, 2013; Raffestin, 1980; Bridge, 2009; De Gregori, 1987; Zimmerman, 1951). Starting from the law allows us to grasp the geo-history of Chilean lithium and its regulation.

The Law: A Social Institution Constrained by and Producer of Territories

As in many countries, the exploitation of underground resources in Chile is based on a concession mechanism. This constitutes both a legal norm and a spatial form that serves to delimit the portion of space within which the exploitation of certain underground resources is permitted. It specifies the conditions of validity in time and space (activities, investments, payment of taxes, etc.). "*[T]he concession model provides the basis for a contractualization between the State and the user that sets the spatial limits and conditions of their activity*" (Nadaï Labussière, 2014:208). In the Chilean case, the "*mining concession obliges its proprietor to develop the activity necessary to satisfy the public interest to justify its granting*" (Art. 24, Constitution of 1980).[4] The fact that the granting of a concession legally responds to a public interest allows the State to organize the area into a productive structure by hierarchizing activities. Thus, the law favors the extraction of subsoil resources over any other form of activity and/or land tenure (Art. 1, 1983 Mining Law). The

[4] Our translation of: "*La concesión minera obliga al dueño a desarrollar la actividad necesaria para satisfacer el interés público que justifica su otorgamiento.*"

legal differentiation of surface and underground leads to a physical and symbolic enclosure of these spatial layers. Consequently, the landowner cannot object to the concessionaire's use and exploitation of surface land achieved through easements, but it is entitle to compensation rights (Bos, 2017; Forgetet al., 2018).

The definition of natural resource rights and the differentiation of legal regimes applicable to the land surface and underground can be analyzed as a geo-legal strategy implemented by Chilean public power. This dominant actor defines the rules and controls space to appropriate the underground resources (Raffestin, 1980). This strategy follows a lucrative objective of *commodification* (Smessaert, 2020). The legal framework for the exploitation of subsoil resources shows that the law is both spatially constrained and productive, as it allows for the creation of new spatial categories that can be superimposed (Bartel et al., 2013). Thus, Chilean law exhibits a strategy for public control over lithium resources, similar to most states, through a legal regime affecting the subsoil that is dissociated from the legal regime affecting the surface.

Lithium Resources Controlled by the State...

As early as 1932, the State, via the mining code, included lithium in the list of minerals that could be exploited under private concessions (without, however, mentioning brines).[5] Nevertheless, it was only in the 1970s that lithium truly became a resource, due to the social utility and market value conferred upon it. The lithium laws of the 1970s and 1980s reflect a geo-legal strategy for asserting state power over the resource and the territory in which it is found.

The legal frameworks that define lithium's status are key factors for understanding the resource's geo-history and resulting actor relations. Through the intermediary of the Ministry of Mining, the executive power declared lithium to be of "nuclear interest" in 1976[6] and of "national interest" in 1979.[7] This legal framework allowed public authorities to remove lithium from the national mining regime, reserving its access

[5] Article 3 of legal decree 488 of the Ministry of Development, 24/08/1932.

[6] Legal decree 1557, Art. 37 (b).

[7] Legal decree 2886.

for potential military and medical uses.[8] In 1964, the resource was placed under the authority of the National Nuclear Energy Commission (CCHEN), which delivers authorizations for exploitation and commercialization.[9] The State's public property and its usufruct on lithium were reinforced during the 1980s, given their inscription at the top of the pyramid of norms:

"The State has the absolute, exclusive, inalienable and imprescriptible ownership of all mines, which include the *covaderas*, metalliferous sands, *salares*, [...] notwithstanding natural and legal persons' ownership over the lands in which they are. Surface lands shall be subject to the obligations and limitations established by law to facilitate the exploration, exploitation, and benefit of said mines" (Art. 24, 1980 Constitution). This public control is reaffirmed by the 1983 mining law,[10] in articles 1 and 7, as lithium cannot be granted in the form of a mining concession to any actor other than the State itself.[11]

The Chilean state's strategy aims to reserve a potentially strategic resource for which demand was low at the time, although several states were projecting its nuclear industrial use (civil, medical, and military) (Forget & Bos, 2022).

...but Produced by Private Mining Companies

Although the resource is legally reserved for the state, Chilean lithium production is nevertheless carried out by two private mining companies: the US *Albemarle* and the Chilean *Sociedad Química y Minera* (SQM) (Colchico, 2020; Forget & Bos, 2022). Production is mainly in the country's northern Atacama region (Fig. 9.2).

[8] "*Por exigirlo el interés nacional, desde la fecha de vigencia de este decreto ley, el litio queda reservado al Estado*" (Legal decree 2886, Art. 3).

[9] Supreme Court decree 432.

[10] Law 18248.

[11] Cf. Article 7: "No son susceptibles de concesión minera los hidrocarburos líquidos o gaseosos, el litio, los yacimientos de cualquier especie existentes en las aguas marítimas sometidas a la jurisdicción nacional ni los yacimientos de cualquier especie situados, en todo o en parte, en zonas que, conforme a la ley, se determinen como de importancia para la seguridad nacional con efectos mineros, sin perjuicio de las concesiones mineras válidamente constituidas con anterioridad a la correspondiente declaración de no concesibilidad o de importancia para la seguridad nacional" (emphasis added).

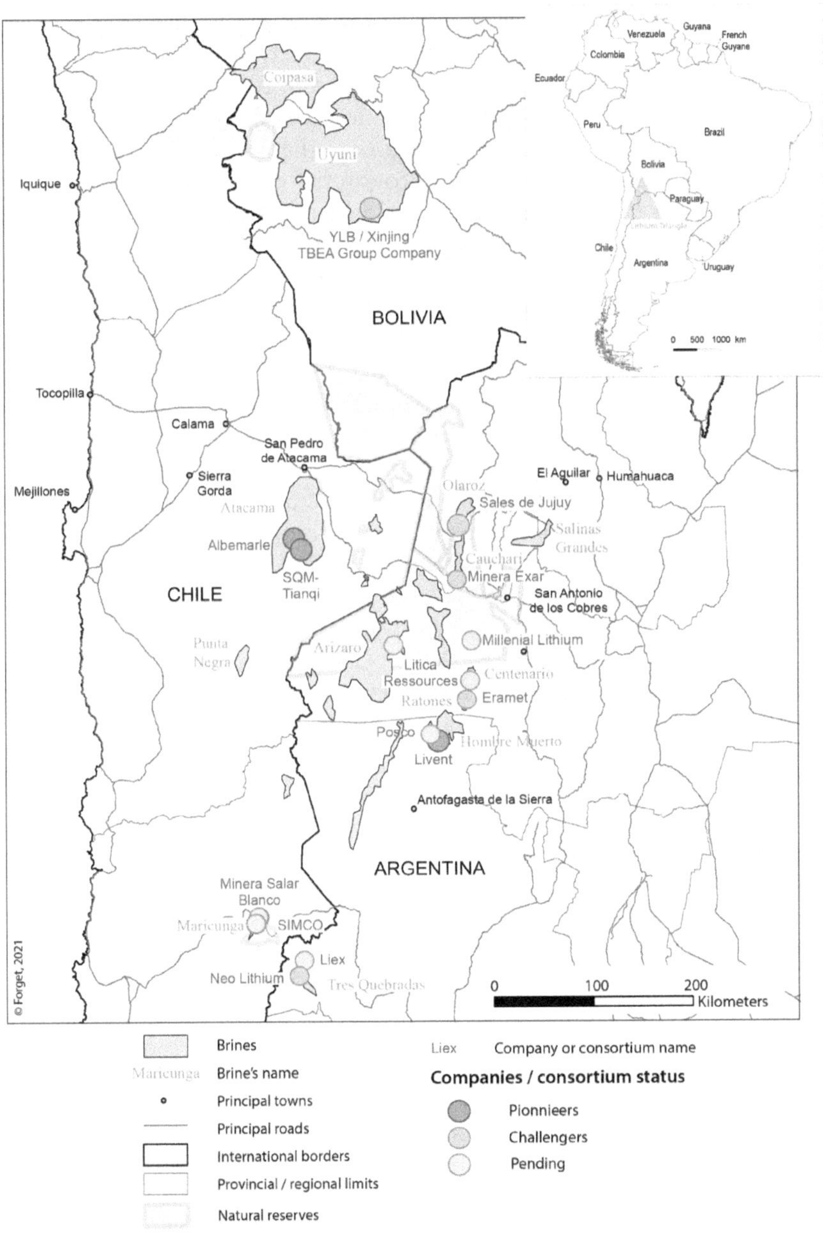

Fig. 9.2 Lithium mining's entrepreneurial landscape in Argentina and Chile. *Source* authors

As lithium laws do not have a retroactive effect, companies that have obtained mining concessions on lithium-rich territories before the rules reserving it for the State can retain their concessions and, therefore, their rights to explore and exploit underground resources, as well as the possibility of contractually transferring them to third parties. This phenomenon is guaranteed by law through the mechanism of administrative concessions.[12] In 1995, *SQM* established a contract with the public institution *Corporación de Fomento de la Producción* (CORFO), which had obtained mining concessions in 1977 on portions of the Atacama Desert. This allowed it to expand its phosphate exploitation activities to include lithium production and chemical processing (OCMAL, OLCA, 2019). A similar strategy of alliance between private and public actors was developed by *Albemarle* (formerly *Rockwood Lithium*, formerly *Sociedad Chilena de Litio*) with this same institution, which allowed it to exploit the resource in the *La Negra* sector, near the city of Antofagasta.[13] Alliances between public and private actors around lithium can be analyzed as a geo-legal strategy used by mining companies to access the *salares*' underground. This strategy is in line with the logic of neoliberal mining codes from the 1980s and 1990s, which promote the direct exploitation of natural resources by private actors at the expense of public sector actors (e.g., Bos, 2015, 2017; Campbell et al., 2005; Magrin, 2013). Nevertheless, after the social revolt of 2019, the state seems to have increased its presence within the lithium sector.

Toward a Return of State Regulation of Lithium Resources

Historically, the Chilean state has controlled the commercialization of lithium through the Chilean Nuclear Energy Commission (CCHEN). This commission can authorize or deny actors the possibility to trade and sets marketable volumes and the duration of these contracts (CCHEN, 2016). In 2021, four companies were involved in lithium trading, which remained dominated by the resource's two main local producers (Table 1).

[12] Articles 7 and 8 of the Mining Law (Law 18248) of 1983.

[13] In the same region of Atacama, two other public mining companies, the National Copper Corporation (CODELCO) and the National Mining Company (ENAMI), hold rights to lithium reserves in the Pedernales and Maricunga *salares* for the former, and in the Aguilar *salar* for the latter, and can transfer these rights to other actors.

Company	Date of authorization	Authorized tonnage (t of LME)	Concession	The *salar* concerned
Sociedad Chilena del Litio[1]	1984	200,000	30 years, extended by 10 years in 2014	Atacama
SQM	1996	180,100[2]	30 years	Atacama
Simbalik Ltda[3]	2011	50,000, reduced to 24,300 in 2013	15 years	Maricunga[4]
Cominor Ingenieria y Proyectos S.A	2013	30,000	15 years	Maricunga

Fig. 9.3 List and conditions of the four companies authorized by CCHEN to market Chilean lithium

The state's decision to grant new permits for marketing resources to other companies in the 2010s appears to demonstrate a desire to increase the volume of exports and intensify production. This strategy was seemingly affirmed in 2019. This phenomenon of new public regulation appears to have provoked a modification in the geo-legal strategies used by *Albemarle* and *SQM*. These strategies are based on differentiating "water" and "brine" resources to ensure the maintenance of their activities and profits but also to avoid the potential legislative return of public authorities to this arena, all while reassuring investors of "good governance" over the environment (Fig. 9.3).

9.3 The Production of Lithium-Bearing Spaces: The Legal Distinction Between Water and Brines

The social construction of lithium also relies on a legal distinction between "water," "brine," and "lithium" resources, as well as the strategic use of this differentiation by private companies in discourse.

The Legal Distinction Between Water, Brines, and Lithium

Throughout the twentieth and twenty-first centuries, various legislative texts have distinguished between water, brines, and lithium. Firstly, the three water codes promulgated in 1951, 1967, and 1981 do not mention brines. The texts describe water in very simple ways by focusing on the distinction between surface waters and groundwaters and between stagnant and running water. While they refer to several types of stagnant water, both natural and artificial, these differentiations do not apply to underground waters. All three codes define the latter very succinctly: "*Groundwater is water hidden in the Earth's core, and which has not been revealed.*"[14] This legal definition reveals a homogeneous representation of hydrogeological systems, preventing one from thinking about and recognizing groundwater's circulation and its diversity, notably the quality and density of the mineral substances it contains. The only legal mentions that recognize variation in groundwater quality refer to mineral waters (1951 and 1967).

By contrast, unlike with mining exploitation, legal texts are late to mention brines. The term *salmuera* (brine) appears in three decrees published by the Ministry of Mining since 1990.[15] In these decrees, no entry refers to its protection or water currents.[16] This late inclusion of brines under the law, as well as their absence from water codes, produces a legal dissociation between lithium, brine, and water. The inclusions of lithium in the law, which precede and are dissociated from those regarding brines, thus contribute to the social construction of brines as a mining resource and not as a water resource. Moreover, mentions of lithium in legal texts are always associated with the productive usage of the resource. The absence of any reference to brines in the 1967 and 1981 water codes reinforces this dissociation.

[14] Our translation of: "*Son aguas subterráneas las que están ocultas en el seno de la tierra y no han sido alumbradas.*" Cf. Art. 2 of the 1981 code and Art. 5 of the 1951 and 1967 codes.

[15] No text mentions its synonyms ("*aguasal*" or "*salazón*").

[16] Cf. Article 2 of decree 209 from 20 December 1990: "*Reglamenta forma de costear el valor de adquisición de las pertenencias mineras, ministerio de minería, el cálculo del costo de pertenencia será determinado por cada contribuyente*". Articles 8 et 9 du décret 64 du 26/10/2017: "*Establece requisitos y condiciones del contrato especial de operación para la exploración, explotación y beneficio de yacimientos de litio en el salar de maricunga y sus alrededores, ubicado en la región de atacama, que el estado de Chile suscribirá con salar de Maricunga*". Article 4 du décret 246 du 27/12/2006: "*Otorga concesión de explotación de energía geotérmica a la empresa geotérmica del norte s.a., en el área denominada el tatio, ubicada en la segunda región de Antofagasta, provincia el loa, comuna de Calama, en virtud del artículo 14 de la ley Nº19.657.*"

This historical construction of the three legal definitions has led to brines' subtraction from the category of water resources, contributing to the *doxa* according to which lithium constitutes a nonmetallic mineral intended solely for productive purposes. This phenomenon entails the development of a local or even global imaginary about the availability of lithium as a raw underground material in an isolated manner, dissociated from water resources. This imaginary, produced from legal texts, is mobilized by the State and companies to deny the negative externalities of this resource's commercial exploitation and to legitimize the opening of new productive sites in the country.

The Discursive Reproduction of Legal Distinction: A Geo-Legal Strategy Used by Mining Companies

This legal distinction is used strategically by companies to deploy their activities through two joint processes. On the one hand, it allows companies to exempt brine from rules governing water extraction, as defined by the 1981 code. Near the Atacama *salar*, the companies *SQM* and *Albemarle* hold brine extraction rights of 1700 l/s and 442 l/s and freshwater extraction rights of 442 l/s and 24 l/s (Jerez et al., 2021), respectively – that is, levels between four and 18 times lower. Besides these lithium-bearing companies, there are other mining enterprises present. A little further north, the mining companies *BHP* and *Minera Zaldivar* hold freshwater extraction rights of 1815 l/s and 625 l/s (Garcés & Álvarez, 2020). In other words, four companies accumulate more than 2900 l/s of freshwater to carry out their activities where the quality and availability of groundwater have not been documented. This phenomenon is exacerbated by the fact that the water code grants mining companies an unlimited right to use waters within the concession area necessary for their production.[17]

However, since the 2000s, the basin has been experiencing a water deficit, with losses exceeding water inflows (Garcés & Álvarez, 2020). Between 2000 and 2015, water consumption outpaced inflows by approximately 1.5 l/s (Amphos 21, 2018), surpassing the basin's water recharge rate. While the General Water Board (DGA), the state agency in charge of regulating water resources, has implemented water use restrictions over

[17] Articles 54, 67, and 56 of the three water codes, respectively.

the past decade, these do not apply to brine extraction. This situation is in line with a liberal legal framework that prioritizes private property rights over water rights (cf. Constitution of 1980 and Water Code of 1981). Until 2018, the State could only halt the granting of new extraction rights on an aquifer but lacked the authority to reduce the levels of extraction flows associated with rights already granted. In addition, while the "Declarations of Prohibition," which aim to stop the granting of new rights in a basin, and the "Declarations of Restriction," which allow the DGA to only grant new rights on a provisional basis, have been applied to specific portions of the basin (Babidge et al., 2019), zoning within these declarations fails to account for the interdependence and circulation between surface and groundwater and, therefore, significantly reduces its effect on water overexploitation.

By contrast, the legal subtraction of brines from water allows companies to construct a positive production narrative that portrays extraction as compatible with economic growth and environmental protection. Lithium companies present brines as too saline for other human activities (especially agriculture), implying that the use of brine to produce lithium would not deprive any other water users. This entrepreneurial discourse, therefore, contributes to the social construction of "lithium" as a mining resource and legitimizes its exploitation.

The legal distinction between water and brines is fully involved in the production of lithium-bearing space. Since 1997, the industry has expanded its area of operations from 20 to 80 km^2 (Garcés & Álvarez, 2020). This expansion continues to be supported by the state, as mentioned above, which wants to strengthen the sector and created the Committee for Nonmetallic Mining Activity[18] in 2016. It also authorizes exploration and exploitation in other regions of the country, notably in

[18] The Corfo Council's agreement no. 2892 of 2015, executed by resolution (A) no. 19 of 2016, aimed to: *"formular un marco normativo e institucional que asegure que el Estado defina las condiciones y participe en la actividad de extracción del litio y de otros minerales relacionados con él, como potasio, boro y magnesio, contribuyendo a dinamizar la exploración y explotación de estos minerales"*

the Maricunga *salar*.[19] The incorporation into law of the different porosities between surface and underground with a reversal of strata tends to destratify the geological strata to recover industrial elements.

9.4 Local Populations' (Geo-Legal) Resistance Tactics

Indigenous[20] communities bordering the Atacama Desert's *salar* have responded in various ways to their incorporation into the development processes and deployed a range of tactics to confront extractivism including geo-legal tactics.

Negotiating and Rebuilding the Links Between Soil and Subsoil

Relationships between local communities and companies are marked by asymmetrical power relations that oscillate between cooperation and association, on the one hand, and mistrust, on the other. At first, these relations are extended informally, then gradually become structured within the CSR (corporate social responsibility) policies and departments dedicated to community relations. These self-regulating practices encourage ethical conduct in entrepreneurial activities, particularly in managing the social and environmental impacts of their operations, under the growing pressure of civil societies. Relationships between these two types of actors

[19] Decree 64, 1 March 2018, Ministry of Mining: "Establece requisitos y condiciones del contrato especial de operación para la exploración, explotación y beneficio de yacimientos de litio en el salar de Maricunga y sus alrededores, ubicado en la región de Atacama, que el estado de chile suscribirá con salar de Maricunga S.P.A"

[20] In Chile, the definition of indigenous communities is based on cumulative laws, including: the 1993 Indigenous Law and the agreement between CONADI (National Corporation for Indigenous Development) and the Ministry of National Assets (MBN), and that of CONADI with the General Water Board (DGA); the definition of Atacameño and Quechua indigenous lands and territories between 1996 and 1999; the creation of indigenous development areas (ADI) in the Atacama *salar* (1997, Article 26 of law no. 19253, approved in October 1995) and the Alto Loa (2003) and, of course, self-identification through sociocultural organizations and the defense of Atacameño groups' interests and their second-level representation, such as the Council of the Atacameño Peoples, the Directing Council of Indigenous Development Areas (Consejo Directivo de las Áreas de Desarrollo Indígena). For an analysis of the processes of indigeneity's strategic mobilization in the context of Chilean mining and the links between multiculturalism and neoliberalism, cf. Gajardo (2020).

are also marked by social conflict. The earliest conflicts arose in 2005–2006, with opposition to *SQM's* extension of mining operations and brine extraction, then in 2007 to requests by the copper mining operation La Escondida (*BHP Billiton*) to pump groundwater. Thus, certain socio-environmental conflicts can be analyzed as a tactic for local renegotiation of the rules to achieve a better balance in the distribution of benefits (e.g., Bebbington et al., 2008; Bebbington & Humphreys Bebbington, 2009; Bos, 2017). These conflicts reveal a new paradigm of relations between lithium-bearing companies and local populations by re-establishing a more horizontal pattern of negotiation. *Albemarle* has engaged in a new model of negotiation with communities based on "shared values" that give communities considerable autonomy, firstly, by associating them with profits generated by the activity, and secondly, by involving them in decisions regarding the use of financial or material resources that result from mining[21] (Gundermann Göbel, 2018).

The populations bordering the *salares* have leveraged the legal framework to position themselves as the principal actors in collective demands for the defense of land, water, and the environment. Within this context, one of the local population's major concerns is the cumulative consumption of water (extraction of brine, water used for production operations, and workers' consumption) and its environmental effects on the quantity of water available for agricultural activities (irrigation and livestock). In addition, they are concerned about the modification of natural landscapes, which are fundamental to these populations' relationship with the world. In these communities' way of life, as elsewhere in the Andes (e.g., Bos & Grieco, 2018; Salazar-Soler, 2004), despite profound ongoing transformations, a strong link unites local populations with natural elements and the territory. *Salares* serve as both cultural and material anchors of local

[21] Specifically, around agreements for direct transfer to the community of Peine by *Rockwood Lithium* and to each of the member communities of the Consejo de Pueblos Atacameños and the indigenous association Consejo de Pueblos Atacameños. "Convenio de cooperación, sustentabilidad y beneficio mutuo entre comunidad indígena atacameña de Peine y Rockwood Litio Ltda.," signed on 8 November 2012 and "Convenio de cooperación, sustentabilidad y beneficio mutuo entre Consejo de Pueblos Atacameños, Comunidad Indígena Atacameña de Rio Grande y otras y Rockwood Litio Ltda.," signed with the Consejo de Pueblos Atacameños and all the communities of the Atacama *salar*, the Área de Desarrollo Indígena Atacama La Grande and the jurisdiction of the Municipio de San Pedro de Atacama, 21 February 2016.

identity; they encapsulate memories of the environment and provide physical support for personal and family biographies (Gundermann & Göbel, 2018).

At the local level, the *salar* communities are organized into indigenous development areas (*Áreas de Desarrollo Indígena* or ADI), territories in which the State promotes harmonious development between indigenous peoples and their communities. The district of *San Pedro de Atacama* contains the ADI of Atacama "La Grande," which seeks to respond to the needs of the 19 *atacameña* communities of the Atacama *salar* region.[22] Some members live in their villages and sectors of origin, but most reside beyond the region, in urban centers. Often, they are wage earners in mining or related companies, employed in the service sector, engaged in small businesses, and frequently travel to their communities of origin. Ownership of the land, natural resources, and collective assets in their territory belongs to the indigenous community. The collective structure is responsible for the distribution of water for irrigation and drinking water for people and animals. It also assumes the group's legal and socio-political representation and its assets toward third parties, hence its involvement in negotiations with public authorities and lithium companies.

The *salar*'s eastern edge corresponds to seasonal grazing areas used by herders from Toconao, Camar, Talabre, Socaire, and Peine during rainy years. However, these areas now contain numerous infrastructures, including energy pipeline networks (Fig. 9.4), roads, mining camps, and other facilities, along with potentially antagonistic socio-economic activities. This overlap of territories and territorialities illustrates the organization of space allowed by national norms that differentiate the surface and the underground.

As a result, access to certain lands and resources (water, fauna) that are socially and culturally important to local people has been altered, disrupting their moral economy and (re)orienting their actions (Scott, 1985). When the villagers of Peine discovered that their village's underground had been registered as mining property, they considered it an "absurdity" given the geophysical and socio-cultural connections between

[22] San Pedro de Atacama, Coyo, Sequitor, Santiago de Río Grande, Machuca, Solor, Socaire, Peine, Quitor, Talabre, Camar, Larache, Catarpe, and Toconao, five of which—Peine, Socaire, Toconao, Talabre, and Camar—are on the edge of the *salar*. All them are between 2400 and 3000 m in altitude and represent no more than 2000 people.

Fig. 9.4 *left*: Freshwater pipelines connecting the SQM extraction site, Atacama; *right*: Entrance to SQM's concession area with flow management infrastructure and, in the background, the closing of the site's access road (Forget, November 2019)

surface-level and underground spaces. To compensate for the environmental externalities of their activities, the State and companies in the area have developed "compensation zones." These environmental conservation projects, under private management, hinder the space's use and disrupt or even prevent the population from performing certain traditional rites that require using the eggs and feathers of pink flamingos and, thus, having free access to the territory and its resources.[23] This phenomenon seems to illustrate a larger process of advancing an extractivist front, now associated with a conservationist front, combining land grabbing, environmental degradation, and social exclusion (Nuñez et al., 2020).

As mentioned, the dissociation of surface and underground compartments is based on their legal separation and a scientific construction of resource flows. Lithium-bearing companies present the underground circulations of aquifers, surface water, and brine as independent and unconnected flows. In this context, publicizing hydrogeological research in the area could demonstrate the narrow connections between these different strata. This would reconcile the various dimensions of this *continuum* by integrating "expert" knowledge with local experiences (Bos, 2017; Grieco, Salazar-Soler, 2013).[24] Recognizing this *continuum*

[23] Interview with V. Figuerroa, Instituto de Arqueología y Antropología (IAA), Universidad Católica del Norte, San Pedro, November 2019.

[24] The notions of "lay" and "expert" knowledge necessarily refer to relations of power and relations of domination, which is why the terms are placed in quotation marks. The notion of the layman can be questioned: "not because it would deny the specificity of

would also legitimize another, more holistic worldview, closer to those of indigenous populations.

The inhabitants of the *salar* basin have developed and transmitted a culture founded on the complementarity between the region's ecological environments and ecosystems for over 11,000 years. Indeed, its first inhabitants developed strategies for adapting to and utilizing an environment characterized by a lack of water and high salinity (Mendez, 2020; Mendez et al., 2020). Thus, there is a practical and spiritual continuity between the ecological subsystems of the *pampa* (desert or sparsely vegetated plains), the *tolar* (an environment dominated by bushes called *tolas*), the *pajonal* (a cooler area where grasses predominate), and the *panizo* (hilltops where vegetation is absent). However, lithium mining threatens this balance. Several studies have shown that the *salares* are fragile ecosystems destabilized by the extraction of brine and groundwater (Liu & Agusdinata, 2020).

Powers and Counter-Powers, Geo-Legal Tactics of Resistance

There is a growing "*expert knowledge produced by a constellation of actors in scientific, technical and legal fields, forming relationship clusters*"[25] on threats to the local hydrodynamics of the Atacama Desert's ecosystems (López Steinmetz & Bing Fong, 2019; Marazuela et al., 2019; Liu et al., 2019; Babidge, 2016, 2020). Most local indigenous communities have had to hire lawyers, technicians, and other experts to defend their positions and attempt to reduce the power asymmetry between mining companies and the State. This is exemplified by the *Atacameño* People's Council, which devotes more than half of the benefits contracted with *Albemarle* to education to create an indigenous counter-expertise to exogenous "expert" knowledge. Several young people from their community are sent to train in national universities to later provide "expert" knowledge that is as legitimate as that of the companies, corroborating

professional knowledge, but because it implies placing oneself from the point of view of the experts [...] and consists in defending this knowledge in opposition to [...] what it is not, as non-specialized or uninitiated [...] Good (1998) thus denounced the use of the concept of the "layman" as it would reflect the establishment of a relationship of domination between [expert/scientific] knowledge and popular knowledge." (Bos, 2017: 484)

[25] Interview with H. Gudermann, Instituto de Arqueología y Antropología (IAA), Universidad Católica del Norte, San Pedro, November 2019.

local experiences and observations. The objective is to carry out independent monitoring and to fill the information gap, as knowledge about the *salar* and its functioning is incomplete, and the information produced is privatized by the extractive companies. These companies characterize the environment, determine "norms," and monitor them. For example, *SQM*, in its environmental impact assessment, aligns the base level with the historically lowest level, allowing it to pump until it reaches low water levels that should only be reached over a century. Similarly, although a public register of water concessions is available online, Chile's private ownership and water market regime means that concessions have not been monitored by government authorities (Bauer, 2002; Budds, 2013). Indigenous communities' actions aim to reclaim independent control, enabling them to better understand the hydrosocial cycle and guarantee their biological and social survival. By inscribing their claims within the technical domain, they can evade state or private strategies that undermine nonexpert knowledge (Boelens, 2015; Grieco & Salazar-Soler, 2013; Linton & Budds, 2014).

In parallel, with national environmental measures and prior consultation processes, Babidge (2020) shows that since 2017, community leaders have been negotiating their relationships with companies. They have begun to demand more from mining companies, refuse to work with them or become increasingly vocal in their accusations about the shortcomings and repercussions of new state regulations on the environment. In this context, social conflict can serve as a potential "weapon" for renegotiating the rules of local social contracts with the mining sector or even state representatives (Bos, 2017).

9.5 Discussion and Conclusion

In this article, we have shown that the social construction of the lithium resource in Chile is based on a national legal framework that dissociates surface and geological strata, denying circulation and links between the two. The "despatialized" legal distinctions are discursively reproduced by companies to pursue their production activities. These companies deploy a geo-legal strategy based on the subtraction of their activities from the normative framework that regulates groundwater exploitation while discursively minimizing the negative environmental externalities produced.

Local populations prioritize the application of geo-legal tactics to regain control over their territory and curb companies' expansionist activities. These tactics are beginning to bear fruit, particularly through the judicialization of their struggles. Following a complaint filed by the Peine and Camar communities with the Indigenous Advisory Council of the *Atacameño* people, the Environmental Tribunal of Antofagasta judged a compliance plan presented by the *SQM* mining company, in response to an investigation conducted by the Chilean environmental regulatory authority (SMA), as "insufficient." The latter found that the company had overdrawn lithium-rich brine, exceeding its rights and effectively threatening the environment's sustainability. The legal ruling curbed the company's expansion plan. The court stated that its decision was based on a precautionary principle, given the fragility of the Atacama ecosystem and the high level of scientific uncertainty. This phenomenon illustrates, on the one hand, a new plurality within the State's position concerning the lithium issue, with public authorities seeming to oscillate between the role of facilitator for private interests and that of mediator between these and the population. On the other hand, it confirms the weight that extra-entrepreneurial production knowledge holds in regulating the mining sector and the environment and, thus, the organization of space. Notably, the institution declared that *SQM* had no way of proving that the mitigation measures it proposed were capable of containing and reducing or eliminating the negative effects generated by the company's infractions, thus stating that the local communities were right. Our research shows that, rather than an excess of power, a lack of clarity, or legal ambiguity, the production of lithium-bearing spaces comes from an imbrication of geo-legal actions constructed simultaneously by the State, mining companies, and local populations. Each of these actors instrumentalizes national law to defend their interests according to their means and their representations of hydrosystems, which are inseparable from the state of knowledge held about them.

Our research also contributes to the fields of legal geography that deal with socio-environmental uprisings. Recently, much research has combined legal geography with political ecology to analyze the capacity of local organizations to change legal frameworks and enforce the human right to live in a healthy environment, access water, and recognize the rights of nature (Ulloa, 2020; Fladvad et al., 2020). These approaches highlight the vision of nature held by dispossessed, often indigenous, populations and the local embeddedness of the construction of these rights (Bellier, 2014; Kinkaid, 2019). We reveal the existence of local resistances that do not structure themselves within the framework of multilevel social and political mobilization and do not explicitly demand recognition of their cultural and environmental rights. The local populations studied contest the lithium industry's misdemeanors by instrumentalizing the law to their advantage, without occupying public or mediatized space. In this sense, we extend reflections on the judicialization processes of environmental disputes and socio-environmental conflicts. The use of the law by local actors to defend their access to natural resources or to protect the environment is complex and sometimes contradictory. Rather than challenging a normative framework or seeking recourse in the courts, we reveal a diversion of national laws. The populations studied build resistance through the law without challenging the normative framework in force.

Complementing the geo-legal lessons, this case is closely linked to questions of stratification, de-stratification, and re-stratification. *Salar* landscapes are evaporative landscapes in which stratification has been established over geological time by a succession of water weathering and evaporation. This stratification corresponds to the dominant vision of this geological form, which has become the common imaginary of these desert landscapes, now considered "petrified." The prospect of a geo-legal definition of the separation of freshwater and brine in legal texts is clearly aimed at establishing, within a legal framework, the stratification induced by a mining development. However, this stratification can be disrupted by the permanent, albeit rare and intermittent, circulation of water from the high-altitude surrounding volcanoes, which allows for the very slow recharging of the saline aquifers. The exploitation of brines to extract mineral salts induces a form of de-stratification by adding anthropogenic circulation through pumping. The "brine stratum" finds its way to the surface, where it evaporates and solificates, forming a new stratum on the surface from the strata below (Fig. 9.5).

The traditional compartments proposed by geologists can be re-stratified—or, at least, the strata definition softened—with the advent of other direct extraction processes that reinject the brines from which the lithium was extracted. Thus, the circulation and permeability of strata are rehabilitated by new technologies.

In conclusion, the extractive industry stabilizes a lithium-producing stratum by establishing a withdrawal loop on a very extensive hydrological cycle: flow, aquifer, and evaporation. The creation of this stratum calls

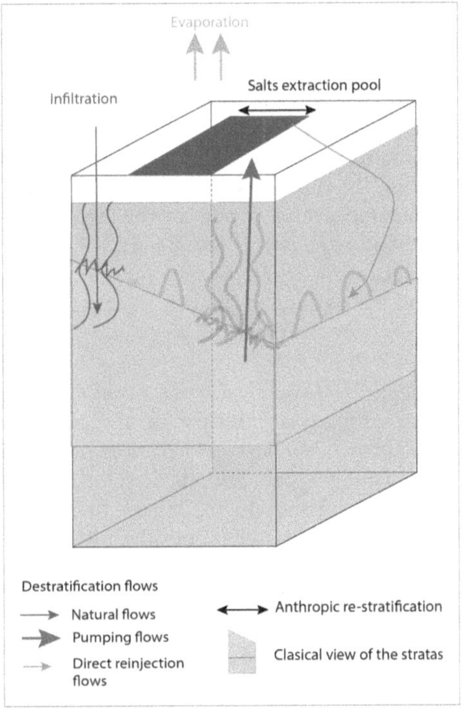

Fig. 9.5 De-re-stratification through natural and anthropic factors

for several operations, both legal (distinguishing brine from aquifer) and sociotechnical (pumping intensity, industrial-scale evaporation), which gradually destabilize certain compartments of the hydrological cycle (loss of high-altitude wetlands, aquifer deficit) to such an extent that the stratum has to pre-empt other environments (seawater conveyance) and find social compromises to continue operating. As a result, this stratification process is eminently disruptive, proceeding by breaking links in several places and fueling de-stratification processes—in the sense of soils and environments becoming less predictable and habitable.

Competing Interests and Acknowledgements This work has benefited from a government grant managed by the Agence Nationale de la Recherche under the "France 2030" program, reference ANR-15-IDEX-0002, via LabEx ITTEM.

References

Amphos 21. (2018). Estudio de modelos hidrogeológicos conceptuales integrados, para los salares de Atacama, Maricunga y Pedernales. Etapa III. *Informe Final. Modelo Hidrogeológico Consolidado Cuenca Salar d'Atacama*. Comité de Minería No Metálica CORFO.

Babidge, S. (2016). Contested value and an ethics of resources: Water, mining and indigenous people in the Atacama desert, Chile. *The Australian Journal of Anthropology, 27*(1), 84–103. https://doi.org/10.1111/taja.12139

Babidge, S. (2019). Sustaining ignorance: The uncertainties of groundwater and its extraction in the Salar de Atacama, northern Chile. *Journal of the Royal Anthropological Institute, 25*(1), 83–102. https://doi.org/10.1111/1467-9655.12965

Babidge, S. (2020). Consultation's overburden: Indigenous participation in the extractives industry in the Salar de Atacama, Chile. *Transformations, 33*. http://www.transformationsjournal.org/wp-content/uploads/2020/02/Trans33_03_babidge.pdf

Babidge, S., Kalazich, F., Prieto, M., & Yager, K. (2019). 'That's the problem with that lake; it changes sides': Mapping extraction and ecological exhaustion in the Atacama. *Journal of Political Ecology, 26*(1), 738–760. https://doi.org/10.2458/v26i1.23169

Bartel, R., Graham, N., Jackson, S., Prior, J. H., Robinson, D. F., Sherval, M., & Williams, S. (2013). Legal geography: An Australian perspective. *Geographical Research, 51*, 339–353. https://doi.org/10.1111/1745-5871.12035

Bauer, C. J. (2002). *Contra la corriente: Privatización, mercados de agua y el Estado en Chile* (1st ed.). TERRAM.

Bauer, C. J. (2004). *Siren song: Chilean water law as a model for international reform*. Resources for the Future. https://doi.org/10.4324/9781936331062

Bebbington, A., Hinojosa, L., Bebbington, D. H., Burneo, M. L., & Warnaars, X. (2008). Contention and ambiguity: Mining and the possibilities of development. *Development and Change, 39*, 887–914. https://doi.org/10.1111/j.1467-7660.2008.00517.x

Bebbington, A., & Humphreys Bebbington, D. (2009). Actores y ambientalismos: Conflictos socioambientales en Perú. *Íconos, 35*, 117–128. https://doi.org/10.17141/iconos.35.2009.371

Bellier, I. (Ed.). (2014). *Terres, territoires, ressources. Politiques, pratiques et droits des peuples autochtones*. L'Harmattan.

Blomley, N. (1994). *Law, space, and the geographies of power*. Guilford Press.

Boelens, R. (2015). Water, power and identity: The cultural politics of water in the Andes. *Routledge*. https://doi.org/10.4324/9781315867557

Boelens, R., Hoogesteger, J., Swyngedouw, E., Vos, J., & Wester, P. (2016). Hydrosocial territories: A political, ecology perspective. *Water International, 41*, 1–14. https://doi.org/10.1080/02508060.2016.1134898

Bos, V. (2015). De l'Etat facilitateur à l'Etat médiateur ? Analyse du positionnement de l'Etat péruvien face aux évolutions du secteur minier depuis les années 1990. In M. Redon et al. (Eds.), *Ressources mondialisées. Essais de géographie politique* (pp. 281–303). Publications de la Sorbonne.
Bos, V. (2017). *Construction sociale de la ressource et renégociation des régulations, Analyse du secteur minier Péruvien* (Doctoral dissertation). Université Sorbonne Nouvelle - Paris 3.
Bos, V., & Grieco, K. (2018). L'eau: ressource naturelle, ressource politique ? *Caravelle, 111*, 59–78. https://doi.org/10.4000/caravelle.3735
Bottaro, L., & Sola Álvarez, M. (2016). La politisation des glaciers en Argentine: une analyse de l'application de la loi nationale sur les glaciers (2010). *Cahiers des Amériques Latines, 82*, 113–128. https://doi.org/10.4000/cal.4353
Braverman, I., Nicholas, K., Blomley, D. D., & Kedar, A. (Eds.). (2014). *The expanding spaces of law: a timely legal geography*. Stanford University Press.
Bridge, G. (2009). Material worlds: Natural resources, resource geography and the material economy. *Geography Compass, 3*(3), 1217–1244. https://doi.org/10.1111/j.1749-8198.2009.00233.x
Bridge, G., Bouzarovski, S., Bradshaw, M., & Eyre, N. (2013). Geographies of energy transition: Space, place and the low-carbon economy. *Energy Policy, 53*, 331–340. https://doi.org/10.1016/j.enpol.2012.10.066
Budds, J. (2013). Water, power, and the production of neoliberalism in Chile, 1973–2005. *Environment and Planning D: Society and Space, 31*, 301–318. https://doi.org/10.1068/d9511
Campbell, B., Hatcher, P., & Lafortune, A. (2005). Nouveaux codes – nouvelles normes. Analyse comparative de la libéralisation des codes miniers en Afrique. In M. Rioux (Ed.), *Globalisation et pouvoir des entreprises* (pp. 69–92). Athéna Éditions.
Colchico. (2020). *Lithium market and Chilean production*. https://www.cochilco.cl/Presentaciones%20Ingls/Lithium%20Market%20a%20Chilean%20Production.pdf
De Gregori, T. R. (1987). Resources are not; they become: An institutional theory. *Journal of Economic Issues, 21*(3), 1241–1263. https://doi.org/10.1080/00213624.1987.11504702
Delaney, D. (2001). Making nature/marking humans: Law as a site of (cultural) production. *Annals of the Association of American Geographers, 91*(3), 487–503. https://doi.org/10.1111/0004-5608.00255
Forget M., & Bos V. (2022). Harvesting lithium and sun in the Andes: Exploring energy justice and the new materialities of energy transitions. *Energy Research & Social Science, 87*. https://doi.org/10.1016/j.erss.2021.102477
Fladvad, B., Klepp, S., & Dünckmann, F. (2020). Struggling against land loss: Environmental (in) justice and the geography of emerging rights. *Geoforum, 117*, 80–89. https://doi.org/10.1016/j.geoforum.2020.09.007

Forget, M., & Carrizo, S. C. (2016). Territorial co-constructions facing the emergence of mining activity in the Argentina Northwest. *Cahiers des Amériques Latines, 82*(2), 31–48. https://doi.org/10.4000/cal.4331
Forget, M., Guibert, M., & Carrizo, S. (2018). La territorialisation en Argentine des activités tournées vers l'exportation (mine, hydrocarbures, agriculture): entre conflits et co-production socio-spatiale. *Les Cahiers d'Outre-Mer, 278*, 249–268. https://doi.org/10.4000/com.9095
Gajardo, A. (2020). Performing the 'India Permitida': The counter-gift of Indigenous women targeted by a corporate social responsibility programme (Chile). *Bulletin of Latin American Research, 40*, 172–187. https://doi.org/10.1111/blar.13143
Garcés, I., & Álvarez, G. (2020). Water mining and extractivism of the Salar de Atacama, Chile. *WIT Transactions on Ecology and the Environment*, *245*. WIT Press. https://doi.org/10.2495/EID200181
Grieco, K., & Salazar-Soler, C. (2013). Les enjeux techniques et politiques dans la gestion et le contrôle de l'eau: Le cas du projet Minas Conga au nord du Pérou. *Autrepart, n°65*, 151–168. https://doi.org/10.3917/autr.065.0151
Gundermann, H., & Göbel, B. (2018). Comunidades indígenas, empresas del litio y sus relaciones en el Salar de Atacama. *Chungara Revista de Antropología Chilena, 50*(3), 471–486. https://doi.org/10.4067/S0717-73562018005001602
Hoogeveen, D. (2015). Sub-surface property, free-entry mineral staking and settler colonialism in Canada. *Antipode, 47*(1), 121–138. https://doi.org/10.1111/anti.12095
Jerez, B., Garcés, I., & Torres, R. (2021). Lithium extractivism and water injustices in the Salar de Atacama, Chile: The colonial shadow of green electromobility. *Political Geography*, *87*. https://doi.org/10.1016/j.polgeo.2021.102382
Kinkaid, E. (2019). "Rights of nature" in translation: Assemblage geographies, boundary objects, and translocal social movements. *Transactions of the Institute of British Geographers*, *44*(3), 555–570. https://doi.org/10.1111/tran.12303
Linton, J., & Budds, J. (2014). The hydrosocial cycle: Defining and mobilizing a relational-dialectical approach to water. *Geoforum, 57*, 170–180.
Labussière, O., & Nadaï, A. (2020). Les temporalités entremêlées des processus de transition énergétique. *Socio-anthropologie, 42*, 93–106.
Liu, W., Agusdinata, D., & Myint, S. W. (2019). Spatiotemporal patterns of lithium mining and environmental degradation in the Atacama Salt Flat, Chile. *International Journal of Application of Earth Observation Geoinformation, 80*, 145–156. https://doi.org/10.1016/j.jag.2019.04.016
Liu, W., & Agusdinata, D. B. (2020). Interdependencies of lithium mining and communities sustainability in Salar de Atacama, Chile. *Journal of*

Cleaner Production, 260, 1208–1238. https://doi.org/10.1016/j.jclepro.2020.120838

López Steinmetz, R. L., & Bing Fong, S. (2019). Water legislation in the context of lithium mining in Argentina. *Resources Policy*. https://doi.org/10.1016/j.resourpol.2019.101510

Magrin, G. (2013). *Voyage en Afrique rentière. Une lecture géographique des trajectoires du développement*. Publications de la Sorbonne.

Marazuela, M. A., Vazquez-Sunee, E., Ayora, C., García-Gil, A., & Palma, T. (2019). The effect of brine pumping on the natural hydrodynamics of the Salar de Atacama: The damping capacity of salt flats. *Science of the Total Environment, 654*, 1118–1131. https://doi.org/10.1016/j.scitotenv.2018.11.196

Marusek, S. (2021). Law and the kinetic environment: Regulating dynamic landscapes. *Scopus*. https://doi.org/10.4324/9781315309378

Melé, P., Merlinsky, G., & Scharager, A. (2020). Les territoires du droit de l'environnement: Effets de la judiciarisation dans le bassin Matanza-Riachuelo (Buenos Aires). *Annales de géographie, 733–734*(3), 87–111. https://doi.org/10.3917/ag.733.0087

Mendez, M. (2020). Genealogy of mining territories in the Atacama Desert: The production of modern waterscapes in Tarapacá region, northern Chile (1853–1924). *The Extractive Industries and Society*. https://doi.org/10.1016/j.exis.2020.05.003

Mendez, M., Prieto, M., & Godoy, M. (2020). Production of subterranean resources in the Atacama Desert: 19th and early 20th century mining/water extraction in the Taltal district, northern Chile. *Political Geography*. https://doi.org/10.1016/j.polgeo.2020.102194

Merveille, N. (2011), Déferlante d'enclosures au Pérou : une nouvelle vague d'altération du lien social et des rapports Nature-Cultures. *Droit et cultures, 62*, 177–193. http://journals.openedition.org/droitcultures/2731

Nadaï, A., & van der Horst, D. (2010). Landscapes of energies. *Landscape Research, 35*(2), 143–155. https://doi.org/10.1080/01426390903557543

Nadaï, A., & Labussière, O. (2014). Recomposer la mer pour devenir offshore : le projet éolien de Veulettes-sur-Mer. *Natures Sciences Sociétés, 22*(3), 204–218. https://doi.org/10.1051/nss/2014039

Nicolas-Artero, C. (2019). *Appropriation de l'eau et production de l'espace. La dimension spatiale du droit de l'eau au Chili* (Doctoral dissertation). Université Sorbonne Nouvelle Paris 3.

Nicolas-Artero, C. (2020). Les tactiques géolégales des organisations d'usagers locales contre l'appropriation de l'eau par les secteurs extractifs au Chili. *Annales de géographie, 735*, 55–80. https://doi.org/10.3917/ag.735.0055

Nuñez, A., Benwell, M. C., & Aliste, E. (2020). Interrogating green discourses in Patagonia-Aysén (Chile): Green grabbing and eco-extractivism as a new strategy of capitalism? *Geographical Review*. https://doi.org/10.1080/00167428.2020.1798764

Olivetti, E. A., Ceder, G., Gaustad, G. G., & Fu, X. (2017). Lithium-ion battery supply chain considerations: Analysis of potential bottlenecks in critical metals. *Joule, 1*(2), 229–243. https://doi.org/10.1016/j.joule.2017.08.019

Raffestin, C. (1980). *Pour une géographie du pouvoir.* Librairies techniques.

Salazar-Soler, C. (2002). *Anthropologie des mineurs des Andes. Dans les entrailles de la terre.* L'Harmattan.

Santoire, E., Desroche, J., & Garcier, R. (2020). Quelles méthodes d'enquête pour les recherches géo-légales ? Retour d'expérience à partir de la mise en concurrence des concessions hydroélectriques en France. *Annales de géographie*, 228–249. https://doi.org/10.3917/ag.733.0228

Scott, J. C. (1985). *Weapons of the weak: Everyday forms of peasant resistance.* Yale University Press.

Smessaert, J., Missemer, A., & Levrel, H. (2020). The commodification of nature, a review in social sciences. *Ecological Economics, 172.* https://doi.org/10.1016/j.ecolecon.2020.106624

Sun, X., Han, H., Fuquan, Z., & Zongwei, L. (2017). Tracing global lithium flow: A trade-linked material flow analysis. *Resources, Conservation and Recycling, 124*, 50–61. https://doi.org/10.1016/j.resconrec.2017.04.012

Toumbourou, T., Muhdar, M., Werner, T., & Bebbington, A. (2020). Political ecologies of the post-mining landscape: Activism, resistance, and legal struggles over Kalimantan's coal mines. *Energy Research & Social Science, 65*, Article 101476. https://doi.org/10.1016/j.erss.2020.101476

Turton, D. J. (2015). Unconventional gas in Australia: Towards a legal geography. *Geographical Research, 53*(1), 53–67. https://doi.org/10.1111/1745-5871.12101

Ulloa, A. (2020). The rights of the Wayúu people and water in the context of mining in La Guajira, Colombia: Demands of relational water justice. *Human Geography (United Kingdom), 13*(1), 6–15. Scopus. https://doi.org/10.1177/1942778620910894

U.S. Geological Survey (USGS). (2020). Mineral commodity. Summaries 2020. *U.S Geological Survey.* https://pubs.usgs.gov/periodicals/mcs2020/mcs2020.pdf

Velez-Torres, I. (2014). Governmental extractivism in Colombia: Legislation, securitization and the local settings of mining control. *Political Geography, 38*, 68–78. https://doi.org/10.1016/j.polgeo.2013.11.008

Zimmerman, E. W. (1951). *World resources and industries. A functional appraisal of the availability of agricultural and industrial resources* (2nd ed.). Harper.

CHAPTER 10

The Origins of Organic Farming: Knowledge, Ignorance, and Practices of "Living Soils" (France, 1930s–1980s)

Céline Pessis

List of Abbreviations

Afes	Association française pour l'étude du sol
Afran	Association française pour la recherche d'une alimentation normale
Inra	Institut national de la recherche agronomique
APPCA	Assemblée Permanente des Présidents de Chambres d'Agriculture
Cneema	Centre national d'études et d'expérimentation de machinisme agricole
OECE/OEEC	Organization for European Economic Cooperation

C. Pessis (✉)
AgroParisTech, SADAPT, Université Paris-Saclay, Palaiseau, France
e-mail: celine.pessis@agroparistech.fr

© The Author(s), under exclusive license to Springer Nature Switzerland AG 2025

O. Labussière et al. (eds.), *Back to the Ground*,
https://doi.org/10.1007/978-3-031-88888-5_10

Fig. 10.1 Descrambe wine label designed by Reiser (1974)

In the early 1970s, young Gérard Descrambe decided against a career in communications and instead planned to take over his father's ailing organic vineyard in Saint Emilion with his brother Christian. Gérard was an avid reader of the satirical magazine *Charlie-Hebdo*, one of the few post-May'68 movement magazines interested in political ecology, and decided to write to the editorial team. One of the founders and key contributors to the magazine, Professor Choron, ordered a case of his wine, then another. Descrambe wine quickly became the editorial team's favorite. Over 30 years, artists such as Reiser (1974) (Fig. 10.1), Wolinski (1974), Gébé (1975), Carali (1976), and Cavanna (1977) have gracefully illustrated some of Gérard's wine labels (Descrambe & Hutin, 2018).

From the outset of this partnership, the vineyard was revitalized and found a new clientele, including consumers at organic trade fairs such as Vinibio, where I met Gérard in 2017. Fifty years later, having overcome many difficulties, Domaine Descrambe is still thriving.

This is a pleasant story, but it may also lead us to overlook how much *Charlie* (especially Cavanna) embraced a scientistic worldview (Pessis, 2014). Yet, it tells us what we know best about organic farming: its transformation after 1968, when it pragmatically met the alternative left.

Historians writing on French organic farming generally concentrate on its political trajectory—often without nuance (Nicolas, 2016; Vrignon, 2017; Bivar, 2018; Trespeuch-Berthelot, 2018). By focusing on a few figures at the national level, they narrate the transition of organic farming from the political right to the left after'68, influenced by young, politicized men and women who started farming, opened up to urban consumers, and engaged with the world of science. The organization *Nature et Progrès* (Nature and Progress) embodied this shift in the 1970s (Woos, 2007; Vrignon, 2021).

The period preceding the environmentalization of the 1970s is thus described in contrast as politically reactionary and influenced by esoteric and anti-science currents.

While Raoul Lemaire's political culture is unquestionable, generalizing this observation to the origins of agrobiology as a whole is problematic and obscures other precursors of agrobiology that have not been adequately studied. Furthermore, the studies that justify such an account have not thoroughly investigated the agronomic knowledge and farming practices that are arguably at the heart of what organic farming is all about.

To make the story more accurate, we need to start with the Descrambe family in the 1950s. Gérard's father was a small winegrower, a member of the socialist party (SFIO) before the war, an atheist, and a Freemason who refused the chemical treatments of the vine promoted in the post-war period. In 1954, he joined the *Comité pour l'humus* (Humus Committee) founded by agronomists to promote organic fertilizers. There he met Michel Binder, an Alsatian agricultural engineer trained in biodynamics. Together, they founded a composting plant, using grape marc from the farm, manure from the nearby stud farm, and other local resources like grape seed flour and horsetail and nettle purins to activate fermentation. The Viter-Humus brochure refuted the "dogma of soluble mineral fertilizers" and claimed to develop an intense soil life that ensured greater resistance of the plants to diseases and parasites. Small winegrowers, as well as large tobacco unions, became regular customers of Viter-Humus (Binder, 1957, p. 2).

Incorporating perspectives from the environmental history of soils and alternative agricultures (Gaynor, 2002; Uekötter, 2006; Vogt, 2007; Treitel, 2017; Erley, 2021), this chapter takes seriously the cognitive dimensions, scientific networks, and agricultural practices that are at the core of organic farming. It explores its origins in France by localizing it in rich and complex, yet marginalized, environmental knowledge and practices, and in the main agricultural debates of the twentieth century. It highlights the ongoing controversies in different scientific communities, the concerns and warnings about soil degradation that run through the agricultural profession, and the development of alternative ways of agricultural intensification based on "living soil" from the 1930s to state recognition of organic farming in 1980.

I use the term "living soil" while building on historical ecology and environmental history of soils, which analyzed soil devitalization—a rapid artificialization, an inhibition of natural symbioses, and biological dynamism due to the increasing use of chemical inputs and the deepening and intensifying of plowing—and an associated knowledge erosion during the agricultural modernization of the latter half of the twentieth century (Uekötter, 2006; Visser, 2010; Winiwarter, 2014). A "living soil" is, therefore, a concept inspired by biology, microbiology, and pedology. It represents a scientific category that has been historically constructed and marginalized yet is experiencing a revival today.

Today, it is a very fashionable notion. Ecological gardeners and farmers view "living soils" as multispecies worlds that must be cared for (Puig de la Bellacasa, 2015; Krzywoszynska, 2019). "Living soils" are also tools for cleaning up pollution and managing environmental problems (Meulemans, 2020). Finally, they represent a new Eldorado for agrochemicals.[1]

However, the notion of "living soil" is not new. For at least a century, it has involved a diversity of attentions, promises, and sociotechnical projects that I explore in my historical investigation into the genealogy of organic farming.

[1] For example, Monsanto has invested heavily in microbial inputs that promise to maximize symbioses and associations between plants and microorganisms, such as to fix nitrogen or solubilize mineral elements.

What is referred to, studied, and implemented as *living* in cultivated soils between the "Great Depression" of the interwar period and the onset of crisis in agricultural productivism in the 1970s? How and by whom was soil life mobilized for agricultural production during this period?

To answer these questions, this chapter is based on a systematic analysis of the main professional and scientific journals, archives of agricultural modernization institutions, and seldom-used archives of organic farming.

I first examine the development of comprehensive soil knowledge in the context of overproduction and quality concerns in the 1930s. Next, in the post-war era—detached from and oblivious to the soil—I explore the persistent criticisms, in the name of humus, of agricultural motorization (deep plowing, separation between field crops and livestock, etc.). Finally, I study the food and natural health networks that valued "living soil" in food quality and human health during the 1950s and 1960s. I will conclude with some remarks on the significance of practices and knowledge about "living soil" in the emergence of organic agriculture.[2]

10.1 Soil as a "Living Organism" (1930s)

In the 1930s, the study of the biological processes of soils, their unique natural history, and their metabolism was valued in the context of overproduction and quality concerns. French agronomists were inspired by Russian comprehensive soil science, which had previously failed to be introduced in agronomy. L'*Association française pour l'étude du sol* (Afes), the French branch of the International Union of Soil Science, was created in 1934. Its founding members often compared the soil to a "living organism" (Demolon, 1932, p. 3), "a true natural living body" (Oudin, 1937, p. 279), or an "organized being" (Erhart, 1940, p. 11). Recognizing the living nature of the soil enabled it to be established as an independent object of research, opening up new ways to rationalize agricultural production based on local knowledge of the soil.

During this period of economic crisis, the exploration of microbial dynamism was frequently presented as a way to save money. The director of the Machinery Testing Station wrote as follows:

To reduce production costs, it would be advisable to make greater use of the free forces of nature, both atmospheric agents (...) and the

[2] This article reiterates some of the arguments I previously set out in previous publications (Pessis, 2020, 2021a, 2021b).

powerful but underexplored microbial dynamism (to save on fertilizers) and to consider the cultivation equipment [only] as an auxiliary and a corrector of the fanciful forces of nature (Ballu, 1936, p. 72).

Another major contribution of this renewed soil science—in addition to improving mineral fertilization—was to contribute to the definition of "terroirs." In the 1930s, the certification of products by their origin and the delimitation of terroirs were widely debated legal and commercial issues (Delfosse, 2004). According to Albert Demolon, the inspector general of the agricultural stations and considered the father of French soil science, the terroir "results from the association of a particular variety with an environment (in other words, a given soil and climate—a 'cru')" (Demolon, 1932, p. 468). Prefacing the publication of the first national pedological map (drawn up by the Russian Agafonoff), Demolon expressed his delight:

Already, the notion of "terroir" or "cru," as expressed by local designations, finds its justification in the definition of genetic types. (Demolon, 1936, p. XIII)

These soil scientists collaborated with the *Comité national des appellations d'origine*, which was created in 1935 to address the wine crisis and improve product quality (Lafforgue et al., 1936). Beyond vineyards, the *Institut pédologique du Bas Rhin*, a pioneering institution established in 1930, produced large-scale soil maps to improve the quality of wheat and seed potatoes by adapting them to the terroirs (Erhart, 1937).

Drawing on complex environmental knowledge, locally situated and attentive to vernacular practices, the beginnings of soil science promised agriculture concerned with "quality" and not just "high yields" (Demolon, 1932, p. 467). Defining soil as a "living organism," the product of interspecific interactions with its own history and stage of development, was significant in the pre-war scientific world. This explains why the first biodynamic writings that valued "living soil" (Pfeiffer, 1938) were so well received by the major press and seriously considered in agronomic circles (De Passillé, 1938; Demolon, 1946, p. 153).

The shortage of fertilizers during the war also justified a revival of organic fertilization practices. The marginalization of biological and pedological approaches in soil sciences after the war was all the more sudden.

10.2 Concerns for the Soil in Post-War Motorized Agriculture

After the Vichy collaborationist regime, which, beyond its agrarian rhetoric, laid the foundations for agricultural modernization (Lyautey & Bonneuil, 2022), the Fourth Republic (1946–1958) was a period of implementing planning and science-based agriculture. The links between the state, agronomic institutions, supply industries, and the organized agricultural profession were gradually strengthened. It was a key period for accounting for the sociotechnical choices made after the war and for locking in agricultural possibilities.

The *Institut pédologique du Bas-Rhin* and the *Laboratoire des sols* (created by Demolon in 1934 in Versailles) were destroyed by the war. The former was never rebuilt, while the latter was only partially rebuilt at a later date. With the creation of the *Institut national de la recherche agronomique (Inra)* in 1946, agricultural research underwent unprecedented development, but comprehensive and localized knowledge of soils did not seem to be a priority. By the end of the 1950s, according to renowned specialist Philippe Duchaufour, French soil mapping was in a "truly humiliating state of inferiority" compared with other European countries (Duchaufour, 1957, p. 3). There was no more training or recruitment in soil microbiology at Inra.[3]

Yet, as we shall see in the second part of this section, on the fringes of modern agronomy, doubts and warnings about soil degradation permeate the agricultural world. The widespread use of tractors is a cause for concern, and learned societies are established to promote agricultural techniques working with soil life.

The Inert and Standard Soil of Modernizers

In 1953, the new director of the *Laboratoire des sols*, Stéphane Hénin, a specialist in soil physics, observed:

The success achieved by the use of fertilizers seems to have resulted in a decrease in interest in the study of the soil itself (Hénin, 1953, p. 463).

The fundamental approaches to the soil complex, promoted by Demolon, became marginalized with the generalization of synthetic fertilizers. With fertilizers, it seemed possible to disregard the diversity of soils

[3] The only serious training in soil science is for scientists called on to work overseas.

and the local particularities of cultivation. Standard formulas were defined according to "average soils," and rapid soil analyses were spreading. In these analyses, "speed and cost are more important than traditional accuracy" (OECE, 1954, p. 99), as explained by the experts of the Organization for European Economic Cooperation (OEEC).[4] The microbial dynamics of humus, which do not easily lend themselves to standardized measurement, were ignored. Soils were increasingly reduced to their physicochemical properties.

Faced with the fear of depleting humus reserves, soil scientists developed a metrology of the organic matter to be conserved in the soil in order to support agricultural productivity, known as humic assessment (Hénin & Dupuis, 1945). Such an approach understood soil exclusively in terms of its functions of structural stability and support for the assimilation of mineral fertilizers.

Manure is viewed in this new light. It becomes a "humic amendment," easily replaced by "humic substitutes" such as artificial manure, green manures, or straw. Meanwhile, its traditional "nutritive" function is delegated to chemical fertilizers (Barbier, 1949). This change contributes to making the separation of intensive cereal farming and intensive livestock farming acceptable and "rational," starting with the disappearance of animals in regions that specialize in field crops.

Thus, soil was increasingly considered as a static entity, a physical support, and a stock of matter to be maintained and consumed rationally. Soils became inert and standardized culture supports, abstracted from their vitality and local specificities. Physical and chemical approaches are becoming dominant in the soil sciences, which are increasingly linked to the state's strategic orientations, such as land-use planning and fertilization policies.

However, this reduction in scientific attention to the living and dynamic character of soils remained a partial and controversial process. Many farmers and agronomists were concerned about the loss of soil life in scientific knowledge but also the flow of matter in the field. This is particularly evident in discussions around motorization.

[4] On the importance of this reductionism, see also Uekötter (2014); Page (2019).

Motorization and Specialization in Doubt

Motorization was a post-war national planning priority, with tractors rapidly spreading from 44,000 in 1945 to 1 million in the early 1960s (Cneema, 1974). However, soil devitalization raises concerns about the major transformations brought by motorization, as an agricultural expert explained:

Insufficiently diversified agricultural systems, repeated mechanical plowing, and the scarcity of working animals due to the use of tractors precipitate this impoverishment [of soil life] (Goislard, 1953, p. 424).

With the specter of the Dust Bowl looming in France and new UN bodies actively debating soil degradation (Selcer, 2018), worries abound. The expert summarized concerns relating to the simplification of crop rotations (necessary to ensure the profitability of machinery), an increase in the speed and depth of plowing, plowing of permanent grasslands (the "fodder revolution"),[5] and a lack of organic fertilization linked to the decline in animal traction and the specialization between arable and livestock farming regions.[6]

These concerns were popularized by the association *L'Homme et le sol* (Man and Soil), created in 1946, and are reflected in the main agricultural representative bodies.[7] "Soil wear and tear and humus loss are characteristic of highly motorized countries," noted the magazine *Chambres d'agriculture* in 1949 (APPCA, 1949, p. 14).

Upon reactivation after the war, the *Assemblée Permanente des Présidents de Chambres d'Agriculture* (Permanent Assembly of the Presidents of the Chambers of Agriculture) (APPCA)[8] devoted its first investigation to "the reconstitution and improvement of soil fertility" (1950–1951). After reviewing the responses from the departmental chambers, the rapporteur unambiguously concluded that "agricultural soils are in great

[5] The "fodder revolution" recommends breaking grasslands with the plow to allow the cereals that follow to benefit from the "cheap humus" bequeathed by the "traditional" agriculture of yesterday. On this dispute, see André Voisin's actions in C. Pessis, *Les leçons…*, *op. cit.*

[6] For more details, *Ibid.*

[7] In the old and conservative *Société des Agriculteurs de France*, the *Académie d'Agriculture de France*, and *Les Chambres d'Agriculture,* key bodies of the rising agricultural profession.

[8] Connecting the 90+ *Chambres d'Agriculture* of the country headed by elected farmers unionists.

danger" and "it is, therefore, important to compile a list of harmful interventions" (APPCA, 1951, p. 27). Consequently, new surveys were launched to address these critical issues.

Manure was identified as "the most important issue for almost all chambers of agriculture" (APPCA, 1951, p. 27) concerning soil conservation. For example, the Ardennes Chamber of Agriculture (in northeast France) wrote: "The earth is a living being, and manure is of paramount importance" (APPCA, 1951, p. 31). Manure was crucial not only for cattle regions but also for large-scale farming regions that resisted cereal specialization and refuse to abandon their livestock. The defense of manure, composed of straw and animal excrement, represented the alliance between cereal crops and livestock, forming an integrated crop-livestock farming system. The Seine-et-Oise Chamber of Agriculture, representing a fertile and "advanced" department north of Paris, summarized the doubts and questions permeating the agricultural world:

"We don't know whether those who cultivate without manure are heading for disaster." In addition, they advised "[caution] in reducing the number of animals [to] keep alfalfa fields" (APPCA, 1950).

The group of agro-practitioners[9] agreed:

The earth, as an organic and living matter, cannot be treated with mechanical methods alone (...) we must return to the complex notion of the farm, with the animal element alongside the plant, the organic element alongside the chemical element.[10]

In 1954–1955, an APPCA survey focused exclusively on farmyard manure/fertilizers ("engrais de ferme") and their improvement. It criticized the recently published OEEC report, which had devalued manure and advocated the recovery and direct use of slurry (OECE, 1954). Although recent data on national manure production, its fertilizing, and its humic value are lacking, the APPCA survey endeavors to quantify "the enormous wealth that farmyard manure is and remains" (125,000,000 tons produced annually, or 92 billion francs in fertilizing value and 375 billion, including humic value) (APPCA, 1955).

The separation between agriculture and livestock farming faced strong criticism in the mid-1950s. It would require extensive propaganda for

[9] Farmers, graduates of the Institut National Agronomique.

[10] L'entretien de la fertilité du sol. Travaux du groupe des ingénieurs agronomes praticiens. (1949, May). *Cahiers des Ingénieurs Agronomes*, (46), 31–33, p. 33.

mineral fertilizers and the development of precise agronomic tools, such as humic assessment, to persuade farmers in arable regions to specialize and abandon their livestock and manure.

While doubts persisted even within professional organizations, the few microbiologists trained before the war, such as Jean Keilling and Jacques Duché, gathered in small learned societies,[11] working to bring manure out of the oblivion into which it had sunk and promote research into subjects neglected by agronomic research and agricultural techniques working with soil life.

Under their impetus, the *Académie d'Agriculture* (Academy of Agriculture) set up a commission in 1955 to experiment with simplified soil tillage and compare the cost of mechanical operations and the effects of the free activity of the soil microfauna (Ballu, 1955a). To limit "a vicious circle of over-equipment" (in that it is necessary to re-mechanize to compensate for the destruction of natural cycles), the commission encouraged technological alternatives that preserve the life of the soil, such as the Marty truncated disc plow, developed by a farmer to "protect bacteria that fix nitrogen from the air" (Ballu, 1955b, p. 326).

As for the *Comité pour l'humus* (1953), on which Gérard's father was active, it promoted organic fertilizers and the development of composting of household waste. These latter recommendations, including the composting of excrement from cities,[12] were even included in the second modernization plan (1953–1957). Although they were pursuing a vigorous motorization policy, the planning bodies were concerned about "soil exhaustion, which unfortunately is already being felt, even in rich regions."[13]

Until the mid-1950s, some agricultural practitioners and experts carefully considered technical options aimed at maximizing the productive phenomena linked to the vitality of the soil. The presence of animals in field crops, the promotion of tools that respect the life of the soil, organic fertilizers, and good manure-making methods were discussed as possible ways of agricultural intensification. It was only in the second half of the 1950s that these options were discarded and reclassified as marginal.

[11] Such as the *Association pour l'Étude de la Fertilité Vivante des Sols* (Association for the Study of Living Soil Fertility), established in 1951.

[12] See Legrand, Dufour, Soyer, Arbarotti, Higgin, this volume.

[13] Archives nationales. 80AJ57. Rapport général des commissions de production agricole et d'équipement rural, October 1953, p. 83.

10.3 Cultivating Living Soil for Human Health

In the second half of the 1950s, the agricultural profession was increasingly involved in developing agricultural policies and consolidating its ties with the chemical and agricultural machinery industries.

The practices and knowledge about living soils became the prerogative of the natural food, hygiene, and health movements, where the meaning of "living soil" fully included the health issue—a dimension that has sometimes been present, albeit timidly, in the agronomic and agricultural arenas.

L'Association française pour la recherche d'une alimentation normale (Afran, the French Association for the Research of a Normal Diet) represented the most academic branch of these movements. It can be compared to the English Soil Association in that it served as an expanded forum advocating for living soil—a forum comprising medical experts, consumers, fraud inspectors, and members of the *Académie d'Agriculture*. It also enjoyed a significant audience.

The Afran network includes microbiologists Jean Keilling and Jacques Duché, as well as marginalized agricultural service officials like Jean Boucher and André Louis.[14] This robust network of advisors works to establish connections between aspects of reality (soil-food-health) that dominant scientific institutions often ignored. The founding doctors of Afran purchased the Combereau estate, an agricultural and medical demonstration establishment, which aims to "restore normality to the soil–plant-animal-human cycle" (Bas, 1950). Microbial processes are central to the "chains and balances of fertility and health" (Birre, 1959, p. 60–61) that link the worlds of soil, plants, animals, and humans. At the heart of this "microbial hexagon," designed by Jean Keilling, is humus formed by a pile of composted manure (Fig. 10.2).

"Each starting at one end of the health chain,"[15] agronomists and physicians were concerned about the disruption of biological balances, which made plants and animals more susceptible to disease and parasites,

[14] André Louis, an active member of the "humus crusade," wrote the preface to Pfeiffer's second book translated into French (Le visage de la terre. Le paysage, expression de la santé du sol(*The Earth's Face—Landscape and Its Relation to the Health of the Soil*), 1949) and advocated for a "vegetal hygiene" approach to arboriculture. He resigned from his position as director of agricultural services in Charente in 1950 to devote himself to teaching at an agricultural college and to managing his new farm.

[15] La santé du sol. (1956, January). *Sol-Alimentation-Santé*, (1), 5–6, p. 5.

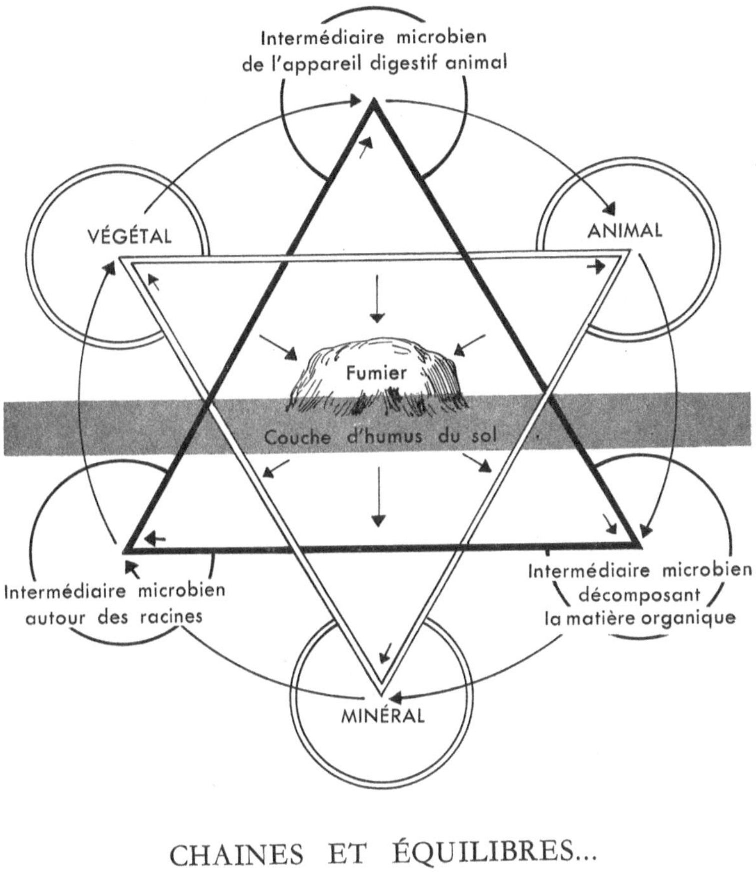

Fig. 10.2 The life star or microbial hexagon (Birre, 1959, p. 60)

and about trace element deficiencies resulting from chemical fertilization. Gathered in learned societies, they proposed to carry out experiments in composting and even to study the vitamins and auxins produced by the fermentation of household waste and manure (Desoutter, 1953).

Young physician Jean-Pierre Ruasse's thesis on bread quality, supported by Afran and nutritionist Jean Trémolières, stressed the importance of soil vitality and humus quality. According to him, the microbial dimension facilitates the transfer of knowledge between soil and the human body. In his essay, Ruasse notes:

> It is not without interest to compare the action of the rhizosphere in plant metabolism with that of the digestive flora in animals, and particularly in humans. We find the same principle of indispensable symbiosis, the same need for a varied population, each element of which seems to have its own action. Will we find the same kind of pathology in humus as in intestinal flora? Let's remind agronomists of the sometimes dramatic accidents of antibiotic digestive therapy (...) Perhaps the doctor's experience can be useful to the agronomist? (Ruasse, 1957, p. 45)

Following this work, Jean-Pierre Ruasse became the director of Afran's quality commission, which proposes to label foods with a high "organic value."

Afran is also engaged in several initiatives in the agricultural world aimed at restoring organic fertility to degraded lands and enhancing the biological quality of the resulting crops. In the Erieux Valley in the Ardèche, the goal is to save a family peach culture, and in Dieulefit in the Drôme, to combat erosion and depopulation in mid-mountain areas. Undertaken by traditional local elites and contrasting with the new technical networks training modern young farmers, these initiatives are forms of resistance to "all chemicals," promoting quality food products.

Beyond its scientific foundation, the defense of a "Living Earth" (special issue of the AFRAN journal) emerges as a method of challenging agricultural dirigisme and policies of motorization and rural exodus. For some, this defense of the freedom and autonomy of the peasantry from agrochemical companies and the state aligns with a politically conservative discourse (Bivar, 2018). For others, however, it was also about supporting, during a time of decolonial movements, "the brutal awakening of the dispossessed peasantry of the world" (Birre, 1959, p. 110).

Outside research institutions, where a physico-chemical approach to soil dominated, an early repertoire of knowledge and practices centered on the care of soil life was being developed. Founded in 1958, the *Groupement des Agriculteurs Biologiques de l'Ouest* (Gabo, the West Organic Farming Group, the first (regional) French organic farming organization) directly inherited the work of the network of scientists around Afran. However, the transformation of "living soils" from an "undone" (but socially asserted) science (Frickel et al., 2010) into a recognized body of theory supporting networks of practitioners organized on a national scale truly took place in the 1960s and especially the 1970s.

10.4 Concluding Remarks About the Beginning of Organic Farming

By the end of the 1960s, new alliances around "living soils" were forged between scientists (e.g., academics, botanists, and young agronomists) and practitioners of organic farming (for example, Messerschmitt, 1974). These collaborations contributed significantly to renewed research and the official recognition of organic farming. Across various institutions, such as some universities, the *Institut Pasteur*, and overseas and forestry research institutes, research on "living soil" entered a new phase, focusing on symbioses, soil–plant associations, and, under the influence of the International Biological Program, soil fauna. Within agronomic research, this resurgence in soil studies prompted tentative institutional changes and early warnings about the environmental impacts of agricultural intensification (Pessis, 2020).

In a climate of social effervescence and criticism of prevailing scientism, these new alliances and shared concerns about soil life sparked growing interest in organic farming in the 1970s, culminating in the French government's recognition of "chemical-free agriculture" in 1980.

This present study reveals that these connections between the conceptions of "living soils," farmers, and scientists have a much deeper history. Here we moved away from a linear history of organic farming, which sees its emergence as recent and associated with advances in knowledge about soils. As Frank Uekötter points out in the German case, the history of soil knowledge is also one of forgetfulness and a fabric of ignorance (Uekötter, 2014). The birth of French organic farming can, therefore, be seen as a response to a persistent and renewed "concern for the soil" in agricultural practices and knowledge, addressing the fabric of ignorance about soil life prevalent in the post-World War II modernizing world. Its emergence is also the result of the practical and discursive work of players who establish connections between aspects of reality (soil-food-health) that dominant specialist knowledge works to disconnect. When established, the first structures of organic agriculture—the GABO (1958), the Lemaire-Boucher method (1964), and the *Nature et Progrès* association (1964)—despite their strategic and political differences, shared a highly coherent technical repertoire. This repertoire was based on principles of "living fertilization," the complementarity between agriculture and livestock farming, and the simplification or "elimination of plowing." As suggested, these technical practices were shaped during debates on agricultural motorization and specialization.

These origins of organic farming can no longer be considered the work of isolated networks, detached from the agricultural worlds or at odds with the realms of science. On the contrary, they are intertwined in the conflicts to define agricultural progress and agronomic modernity, as well as in complex yet marginalized environmental knowledge and practices.

Competing Interests and Acknowledgements I would like to thank the funder of this work, the IFRIS (Institut Francilien Recherche Innovation Société), the participants of the European Society for Environmental History writing group who discussed a longer version of this article, and the Rachel Carson Center, where I presented the work.

Bibliography

Archives and Primary Sources

APPCA. (1949, October-December). La motorisation en agriculture. *Chambres d'agriculture*, (1), *14*.

APPCA. (1950, October-December). Reconstitution et amélioration de la fertilité des sols. *Chambres d'Agriculture*, (5), 17–18.

APPCA. (1951, April-June). Rapport Reconstitution et amélioration de la fertilité des sols. *Chambres d'agriculture*, (7), 26–31.

APPCA. (1955, September 1). La construction des fumières et fosses à purin. *Equipement-Agriculture*, (83), 5.

Archives nationales. 80AJ57. (1953, October). Rapport général des commissions de production agricole et d'équipement rural. p. 83.

Ballu, T. (1936, June). Résumé des travaux du Congrès du travail du sol (2ème section). *Bulletin de l'Afes*, *II*(2), 71–78.

Ballu, T. (1955a, April 2). À propos d'un mémoire sur "la culture sans versoir". *CR Ac. Agri.*, *326*.

Ballu, T. (1955b, December 14). Études sur le travail du sol. *CR Ac. Agri.*, 716–719.

Barbier, G. (1949, June-July). Essai de mise au point sur la question de l'humus. *Bulletin technique d'information des ingénieurs des services agricoles*, (41), 319–327.

Bas, J.-W. (1950, March). Le rétablissement du normal dans le cycle sol, plante, animal, homme. Nécessité d'une démonstration sur un domaine agricole et fermier. *Sol et vitalité*, (1), 19–29.

Binder, M. (1957). *Procédé « Viter-Humus »*. Personal archives of Gérard Descrambe.

Birre, A. (1959). *Un grand problème humain, l'humus*. Organisation scientifique pour l'entretien de la vie.

Duchaufour. (1957). Archives du Groupe d'Etudes des Problèmes de Pédologie Appliquée, Inra Orléans. (1957, October 8). *Rapport de Mr. Duchaufour suite à la Réunion du Groupe de travail de la cartographie des sols de la FAO, Bonn, Nancy*. p. 3.

de Passillé, R. (1938, March 20). Fécondité de la terre. *Le Figaro*, 8.

Goislard, H. (1953, September). La motorisation agricole recèle-t-elle des dangers? *L'agriculture pratique, 117*(9), 423–424.

Hénin, S., & Dupuis, M. (1945). Bilan de la matière organique des sols. *Annales Agronomiques*, (1), 17–29.

Hénin, S. (1953, July-August). Pédologie et fertilisation. *Bulletin Technique d'Information des ingénieurs des services agricoles*, (81), 463–468.

Lafforgue, G., Riedel, C.-E., & Franc de Ferrière, J. (1936). *Les graves de Bordeaux. Relations entre l'évolution et la vocation culturale*. DSA de Gironde.

OECE. (1954). *Fumures minérales et organiques. Progrès réalisables en Europe*.

Oudin, A. (1937). La pédologie. Son objet, ses méthodes, son intérêt agronomique. *Bulletin de l'Afes, 3*(4), 279–293.

Pfeiffer, E. (1938). *Le visage de la terre. Le paysage, expression de la santé du sol*. La Science spirituelle.

Ruasse, J.-P. (1957). *Vrai pain, vraie santé*. Afran.

Secondary Sources

Bivar, V. (2018). *Organic resistance: The struggle over industrial farming in postwar France*. University of North Carolina Press.

Cneema. (1974). *La constitution du parc de tracteurs agricoles depuis 1945*. Cneema.

Delfosse C. (2004). Le pays et ses produits: défense et illustration d'une identité. *Études sociales*, (139–140), 117–138.

Demolon, A. (1932). *La dynamique du sol*. Dunod.

Demolon, A. (1936). Préface. In V. Agafonoff (Ed.), *Les sols de France au point de vue pédologique* (pp. XII–XIII). Dunod.

Demolon, A. (1946). *L'évolution scientifique et l'agriculture française*. Flammarion.

Descrambe, G., & Hutin, D. (2018). *Grand cru déclassé. Appellation Charlie Contrôlée*. Éditions de l'Épure.

Desoutter, J. (1953). Programme de l'Association pour l'Étude de la Fertilité Vivante des Sols. *L'Alimentation Normale,* (3–4), 59–62.

Erhart, H. (1937). *Traité de pédologie. Tome II. Pédologie Agricole.* Institut Pédologique du Bas-Rhin.

Erhart, H. (1940). La pédologie, science des sols. *Annales de l'Institut Technique du bâtiment et des Travaux publics, 5*(1), 12–26.

Erley, M. (2021). *On Russian Soil: Myth and materiality.* Cornell University Press.

Frickel, S., Gibbon, S., Howard, J., Kempner, J., Ottinger, G., & Hess, D. J. (2010). Undone science: Charting social movement and civil society challenges to research agenda setting. *Science, Technology, & Human Values, 35*(4), 444–473.

Gaynor, A. (2002). Looking forward, looking back: Towards an environmental history of salinity and erosion in the eastern wheatbelt of Western Australia. In R. Gaynor, A. Haebich, & J. Trinca (Eds.), *Country: Visions of land and people in Western Australia* (pp. 105–123). Western Australian Museum.

Krzywoszynska, A. (2019). Caring for soil life in the Anthropocene: The role of attentiveness in more-than-human ethics. *Transactions of the Institute of British Geographers, 44*(4), 661–675.

Lyautey, M., & Bonneuil, C. (2022). Les origines allemandes et vichystes de la modernisation agricole française d'après 1945. *Revue d'histoire moderne & contemporaine, 69*(2), 86–113.

Messerschmitt, H. (Ed.). (1974). *Encyclopédie permanente d'agriculture biologique.* Debard.

Meulemans, G. (2020). Urban pedogeneses: The making of city soils from hard surfacing to the urban soil science. *Environmental Humanities, 12*(1), 250–266.

Nicolas, F. (2016). Une croisade morale inachevée. La représentation des mondes agricoles dans le journal agriculture et vie. *Études rurales*, (198), 97–114.

Page, A. (2019). Fertility from urban wastes? The case for composting in Great Britain, 1920s–1960s. *Environment and History, 25*(1), 3–22.

Pessis, C. (Ed.). (2014) *Survivre et vivre. Critique de la science, naissance de l'écologie.* L'Echappée.

Pessis, C. (2020). Histoire des "sols vivants": genèses, projets et oublis d'une catégorie actuelle. *Revue d'anthropologie des connaissances, 14*(4).

Pessis, C. (2021a). De la "croisade pour l'humus" à l' "agriculture biologique". Alertes savantes et mouvements paysans face à la dégradation des sols (1948–1958). In M. Lyautey, L. Humbert, & C. Bonneuil (Eds.), *Les modernisations agricoles au XXe siècle* (pp. 233–247). Presses Universitaires de Rennes.

Pessis, C. (2021b). Les leçons de l'agriculture américaine? Motorisation et souci du sol sous la IVe République. *Le Mouvement social*, (277), 67–82.

Puig de la Bellacasa, M. (2015). Making time for soil: Technoscientific futurity and the pace of care. *Social Studies of Science, 45*(5), 691–716.

Selcer, P. (2018). *The postwar origins of global environment. How the United Nations built Spaceship Earth*. Columbia University Press.

Treitel, C. (2017). *Eating nature in Modern Germany. Food, agriculture and environment, c.1870 to 2000*. Cambridge University Press.

Trespeuch-Berthelot, A. (2018). La genèse de l'agriculture biologique en France et en Allemagne de l'entre-deux-guerres aux années 1970: circulations transnationales et cultures politiques. In O. Hanse, A. Lensing, & B. Metzger (Eds.), *Mission Écologie. Tensions entre conservatisme et progressisme dans une perspective franco-allemande* (pp. 91–114). Peter Lang.

Uekötter, F. (2006, Autumn). Did they know what they were doing? An argument for a knowledge-based approach to the environmental history of twentieth-century agriculture. *German Historical Institute Bulletin, 39*(3), 145–166.

Uekötter, F. (2014). Ignorance is strength: Science-based agriculture and the merits of incomplete knowledge. In F. Uekötter & U. Lübken (Eds.), *Managing the unknown: Essays on environmental ignorance* (pp. 122–139). Berghahn Books.

Visser, J. (2010). *Down to earth. A historical-sociological analysis of the rise and fall of 'industrial' agriculture and of the prospects for the re-rooting of agriculture from the factory to the local farmer and ecology* (Doctoral thesis). Wageningen University.

Vogt, G. (2007). The origin of organic farming. In W. Lockeretz (Ed.), *Organic farming. An international history* (pp. 9–29). CABI.

Vrignon, A. (2017). *La naissance de l'écologie politique en France. Une nébuleuse au cœur des années 68*. PUR.

Vrignon, A. (2021). L'agrobiologie face à la modernisation agricole. Nature et Progrès, de 1964 au début des années 1980. In C. Bonneuil, L. Humbert, & M. Lyautey (Eds.), *Histoire des modernisations agricoles au XXe siècle* (pp. 279–292). Presses Universitaires de Rennes.

Winiwarter, V. (2014). Environmental history of soils. In M. Agnoletti & S. Neri Serneri (Eds.), *The basic environmental history* (pp. 79–119). Springer.

Woos, N. (2007). Un monde agricole insoumis: Agriculture biologique et agrarisme à travers la revue Nature et Progrès (1964–1974). In P. Cornu & J.-L. Mayaud (Eds.), *Au nom de la terre. Agrarisme et agrariens, en France et en Europe, du 19e siècle à nos jours* (pp. 349–360). La Boutique de l'Histoire.

CHAPTER 11

Human Manure as Activism: Composting Excrement as an Alternative Approach to Soil Fertilization

Marine Legrand, Etienne Dufour, Mathilde Soyer, Alexsandro Arbarotti, and Marc Higgin

11.1 Introduction

January 2023. The microphones of Radio France are on the prowl. At the Trou Salé farm, located on the Saclay plateau south of Paris, an employee of the regional chamber of agriculture uses a pump to transfer several cubic meters of human urine from a storage tank to the vehicle that will spread it on the fields. The radio program documents field studies

M. Legrand (✉) · M. Soyer · A. Arbarotti
Ecole Nationale Des Ponts Et Chaussées, Champ-sur-Marne, France
e-mail: marine.legrand@enpc.fr

E. Dufour
Universite Paris 1 Pantheon-Sorbonne, Paris, France

M. Higgin
École Nationale Supérieure D'Architecture De Grenoble, Grenoble, France
e-mail: marchiggin@icloud.com

© The Author(s), under exclusive license to Springer Nature Switzerland AG 2025

O. Labussière et al. (eds.), *Back to the Ground*,
https://doi.org/10.1007/978-3-031-88888-5_11

that have been carried out here on the impact of using human urine as fertilizer on various crops, including wheat and maize.

In recent years, the agronomic potential of human excreta as fertilizer has drawn the attention of various institutions in France. This interest, partly from the scientific community, follows in the footsteps of a social movement known as ecological sanitation, which has been developing for several decades. By redesigning the systems that manage human excreta, the movement questions the relationship between habitat and environment. Although it began in rural areas, it is now beginning to infiltrate the practices of urban development from the margins (Joveniaux et al., 2021; Legrand et al., 2021). The discourses that accompany and promote these alternative practices challenge the conventional sanitation model with its flushing toilets, extensive sewer systems, and treatment plants. Despite the diversity of actors and approaches within this movement, all focus on critiquing the linearity of nutrient flow management, which is intrinsically tied to pollution (Le Noë et al., 2018; Esculier & Barles, 2019). This critique intersects with more general concerns about the sustainability of large networked infrastructure (Tarr & Dupuy, 1988; Coutard, 2010). Authors and activists condemn the material and symbolic concealment performed by centralized sanitation systems concerning the fate of our bodily waste (Hawkins, 2004). They also criticize the production of "modern" housing, which maintains degraded relationships with natural environments. These debates often extend further, linking the narrow view of human excreta as waste with the neglect of soil, its life, and its maintenance within contemporary systems of industrial agricultural production (Legrand, 2020).

More fundamentally, these debates converge around the notion of a "metabolic rift" between the city and the countryside, first conceptualized in the nineteenth century by Marx and his contemporaries. Later explored by eco-Marxist thinkers (Foster, 1999; Saitō, 2017), the "rift" refers to the disruption of metabolic cycles when organic waste from cities is no longer recycled to enrich local soils but instead replaced by the intensive use of synthetic and fossil fertilizers. This shift facilitates regional specialization and agricultural industrialization, integrating agriculture into a global market and contributing to strong political asymmetries between regions. These combined phenomena, which cause serious disruptions in biogeochemical cycles, stem from an industrial sociotechnical system that emerged in the nineteenth century and has been consistently reinforced since (Steffen et al., 2015).

Today, the metabolic rift hypothesis is widely echoed by biogeochemistry experts, who warn that nutrient cycles are among the planetary boundaries at serious risk of being swept away (Esculier, 2018; Esculier & Barles, 2019). Since the early 2000s, these arguments have been used by various scientific disciplines (industrial and territorial ecology, circular economy, etc.), as well as public and private actors, to promote the recycling of urban organic residues, including both biowaste and excreta (Barles, 2022; Kawa et al., 2019).

From a certain point of view, the current interest in composting toilets and the emergence of supply chains in the agricultural sector signal an unexpected revival of interest in human manure. This interest, which had its heyday in the second half of the nineteenth century, nearly disappeared in the twentieth century. Beyond the massive investment in modern centralized sanitation systems, this decline was also due to the rise of the fertilizer industry, both mined (P and K) and synthetic (N), along with the emerging science of agronomy structured and dominated by a reductionist vision—strictly physiological and chemical—of plant growth, often ignoring the complex roles of soil and organic matter. However, we can also discern the emergence of discourses related to the practice of composting toilets in various Western countries from the early twentieth century. These discourses were linked to heterodox approaches in agronomy, where fertility was seen as closely connected to soil biology and the notion of humus.

While the contemporary context is marked by a nearly complete metabolic rift between city and countryside, to what extent do these alternative management practices of human excreta challenge today's conventional—industrial and extractivist—agricultural model? How do they compare with the dynamics of large-scale agricultural valorization of human excreta that existed in the nineteenth century, not only in terms of scale but also in the motivations of the actors involved? Finally, what is the potential transformative impact of these practices on the food system, to which they currently contribute only marginally? This chapter consolidates various ongoing lines of research in the social sciences concerning the management of human excreta. By combining historical and contemporary approaches, it seeks to juxtapose different past and present contexts where centralized sanitation systems are not taken for granted as conditions of housing and inhabiting. Without drawing clear continuities between these contexts, our analysis aims to highlight the practical modalities of returning human waste to the soil as fertilizing material, the scales involved, and the associated motivations of the actors.

This work is based on various research materials.[1] The first part of the analysis derives from an in-depth historical study on the evolution of household waste and sewage management throughout the twentieth century (Dufour, 2024), taking the reader through the golden age of human manure and its subsequent decline and marginalization amid agricultural industrialization. The second part is based on bibliographic research into the emergence of ecological sanitation in the 1960s, supplemented by archival research to identify key historical documents from this movement. It reveals how renewed interest in excreta arose at the margins of the industrial world, through experiments in the 1960s and 1970s within countercultural movements. The third and final parts draw on sociological research into contemporary composting toilet practices in rural France, using a series of semi-structured interviews with domestic dry toilet users. This explores how a relatively small-scale practice is renewing relationships between habitat, food production, and the living environment.

11.2 When the Valorization of Human Excreta was Fully Integrated into Urban/Rural Relations?

In France, the agricultural value of human excreta was intensely studied during the nineteenth century but gradually declined over the following century, except for a resurgence during the Second World War. What was the context in which the large-scale collection, treatment, and distribution of human urine and feces as fertilizer became a useful agricultural strategy? These periods of high agricultural value for human excreta occurred when shortages of fertilizing materials (lack of supply) intersected with increasing agricultural industrialization and growing, mostly urban, populations (high demand), driving the need for scale and infrastructure. The

[1] This chapter is the result of an interdisciplinary discussion among two sociologists, a historian of urban planning, and two anthropologists. The first survey was conducted by Etienne Dufour as part of his thesis on the history of biogeochemical policies in the Île-de-France. The second and third surveys were conducted by Marine Legrand and Mathilde Soyer, respectively, as part of the research project "To the toilets… and after?" on the development of ecological sanitation in France (OCAPI program, www.leesu.fr/ocapi). The intersection of these works benefited from the theoretical contributions of Marc Higgin and Alexsandro Arbarotti.

retrospective that follows is based on a literature review of nineteenth-century developments and a comprehensive review of specialized articles or books from the period of the Second World War.

Was the Nineteenth Century a Golden Age for the Agricultural Use of Sewage?

The nineteenth century represents a pivotal period for the agricultural use of human excreta in France. Building on earlier "traditional" practices, this utilization was supported by hygienists, chemists, and various public intellectuals and activists (writers, social reformers, etc.) during this era of significant industrialization. In the French context, notable figures include Adolphe-Auguste Mille, Jean-Baptiste Dumat, Jean-Baptiste Boussingault, Victor Hugo, Pierre Leroux, and Joseph Déjacques. On the ground, it was organized and implemented by engineers, entrepreneurs, and farmers, sometimes on a very large scale but always with strong territorial roots and intense exchanges between urban and rural areas. This economy has been described in various publications about Paris (Barles, 2005b), Brussels (Kohlbrenner, 2014), and Lyon (Aguerre, 2003; Adler, 2020).

During this period in Europe, urban sewage materials were generally used as fertilizer: either as "muck" through the *Flemish method* (direct spreading) or following industrial transformation into compost or ammonium sulfate. Developed in the late eighteenth century, these fertilizers achieved international success, and manufacturing plants proliferated on the outskirts of major cities. With the gradual and still partial introduction of sewerage systems by the end of the nineteenth century, the authorities maintained the principle of agricultural recycling of excreta was maintained by the authorities, directing wastewater to areas of demand, primarily market gardens on the city outskirts. This was the case for Paris from 1869 and Lyon from 1899. These projects sparked numerous controversies and were not without conflict among doctors, engineers, local populations, and authorities. To promote this system of spreading human muck on productive land, engineers from the "Ponts et Chaussées" (a public body of engineers overseeing public works in France) relied on a strong technical and social network, which prevailed over the opinions of doctors who could not yet rely on the legitimacy of a unified medical and epidemiological science (Barles, 2005a). The principle of "scientific rationality" was widely invoked to shift controversies from the political arena to that of technical expertise (Carnino, 2013). However, human muck spreading did not occur everywhere. In Brussels, a project similar to that of Paris remained pending and never materialized. Instead, the decision was made to sacrifice the city's main river, the Senne, by converting it into a public sewer. This marked the beginning of the gradual abandonment of agricultural recycling of excremental materials in Brussels and elsewhere.

This large-scale use of human manure supported the initial industrialization of French agriculture in the context of fertilizer scarcity. The Haber–Bosch process for the artificial synthesis of ammonia from atmospheric nitrogen was developed only in 1913 and further refined after the First World War (Smil, 2001; Gorman, 2013). Fertilizer sources were already diversifying in the nineteenth century with the use of Chilean guano and the discovery of local and foreign fossil resources. The exploitation and mobilization of these resources were made possible by international trade, colonialism, advances in chemistry, and the growing use of fossil fuels. However, this increased supply was insufficient to meet the rising demand resulting from demographic growth and accompanying urban and industrial development, leading to continued shortages. This prompted the search for alternatives to the limited supply of animal manure, which remained the primary source of soil fertility. Among these alternatives, urban excreta, as either transformed sewage or wastewater rich in organic matter, proved to be a valuable fertilizer. For peri-urban agriculture, it was "vital" (Aguerre, 2003). Although this assessment was not universal, there was relative consensus during the period and a temporary alliance between urban and agricultural actors.

Another "Golden Age": Second World War

With the turn of the twentieth century, this large-scale agricultural use of human excreta began to decline due to the increasing dominance of synthetic fertilizers (Barles, 2005b, 2021). However, the world wars, with their periods of scarcity, were exceptions to this steep downward trend. Human-derived fertilizers experienced a renewed interest in contexts where the scarcity of fertilizing materials was once again felt.

The Second World War, more than the First, led to a significant reduction in resources available for agriculture (Cépède, 1961). While the consumption of synthetic fertilizers had been steadily increasing over the previous three decades, supply was abruptly cut off with the outbreak of hostilities. The conflict led to a collapse of international trade, limiting access to phosphate deposits, primarily in Morocco. Industrial plants using the Haber–Bosch method were repurposed to produce explosives or became targets of enemy bombings. In the French context, potash mines in Alsace fell under German control. For example, the supply of synthetic nitrogen fertilizer available to farmers was reduced by 75% from the spring of 1939 to 1945. The reductions were even more drastic for phosphorus (−94%) and potassium fertilizers (−98%). However, agricultural production decreased by "only" one-third over the same period. In addition to mobilizing land banks, the use of traditional natural sources of fertilizers (manure, green fertilizers, household waste, human excreta) helped to limit the impact of synthetic fertilizer shortages, ensuring continued soil fertility.

Faced with this crippling shortage of fertilizer, a diverse group of actors in agronomy, urban sanitation, planning, and architecture (Puget, 1939) sought and promoted various solutions, among which the mobilization of human manure played a significant role. During the conflict, certain approaches within agronomy (Pessis, 2020) and holistic medicine (Bernard, 2019), which originated in the 1930s advocating soil fertility and nutrition linked to soil life and organic inputs, found a wider audience. The agronomist and "father of pedology" in France, Albert Demolon, enthusiastically promoted the use of human excreta for the production of "artificial manure," based on a method developed before the war at the central agronomy station in Versailles (Demolon & Burgevin, 1937). This method was widely disseminated in the press and specialized journals during the conflict (Burgevin, 1940; Boyer, 1945). Likewise, in their books popularizing agricultural science, chemical engineer Charles Bauer (1942) and agronomist Roger Dusseaulx (1943) advocated for the creation of household waste composts and the recycling of human excreta as solutions to the shortage of synthetic fertilizers. These recommendations were also echoed by the Vichy government, which, like those of other warring countries (England, the United States, Italy, Germany, and Japan), encouraged the recuperation and recycling of all waste to cope with shortages. In particular, the use of human-derived fertilizers was part of a broader policy to promote "family" (or "war") gardens (Chouard, 1942). Finally, the resurgence of epidemics related to the use of non-sanitized excremental materials, such as typhoid and ascariasis, suggests a generalized but unsafe (uncomposted) use of these materials in crops during the dark years (Lacroix, 1943).[2]

While agronomists still ponder whether it is possible to "maintain the fertility of the soil without chemical fertilizers" (Maïs, 1946), the use of these alternatives remained recommended for some time after 1945. In the "orientations and equipment plan" for French agriculture on behalf of the General Confederation of Agriculture, agronomist René Dumont, a future ecologist presidential candidate, proposed to generalize the use of the "human herd" for fertilization. He argued that "human fertilizers—sewage and urine—are the most important of these [...] additional sources of fertilizers; their return to the soil would be as useful as that of (animal) manure, especially in densely populated areas" (Dumont, 1946). In the 1950s, the use of wastewater and human fertilizers was still recommended by some members of the General Planning Commission (Keilling & Casays, 1953) and was even included in its second "modernization and equipment plan." Furthermore, ongoing agronomic experiments highlighted the benefits of sewage for crop production (Desjobert, 1950), alongside innovations to facilitate their agricultural use under sanitary conditions that limited the spread of fecal germs. In this context, Auguste Abdon's "separator seat" is regarded today as a pioneering invention in the field (Abdon, 1948).

These practices persisted for a few years after 1945. While the Second World War slowed their abandonment, it did not stop it (Barles, 2021). In the wake of the so-called "trente glorieuses," the use of human fertilizer in agriculture became marginal. While it persisted in very specific contexts, the State pursued a policy of agricultural intensification that relied heavily on promoting the production and use of industrial fertilizers (Martin, 2017). Their competitors, particularly urban sludge and sewage materials, were marginalized (Dufour, 2023). Moreover, while composts are still produced from household waste, sewage materials gradually disappeared with the generalization of sewerage systems (Dufour, 2024).

As we have seen, the nineteenth century and the period of the Second World War are marked by significant agricultural use of human urine and feces in France, even contributing to a certain industrialization of agriculture (Tomic, 2017). In both instances, these practices emerged from a confluence of agricultural necessity, due to the scarcity of fertilizers of all

[2] It can also be a consequence of declining living conditions, impoverishment of the diet, and therefore the immunity status of populations over the period.

kinds, and the voluntary, organized promotion by scientific, technical, and political institutions. Human excremental materials held economic value and became integral to the fertilizer industry and the broader agricultural economy. However, both its infrastructure and practices have now been relegated to a bygone era.

11.3 A Reinvention from the Margins of the Industrial World

In the second half of the twentieth century, even as sewer systems became more widespread in France, the practice of returning human excreta to the soil did not vanish entirely. What were the contexts that encouraged this persistence? What motivations sustained it? On the one hand, human-derived fertilizer continued to be used in rural areas out of necessity, due to the delayed implementation of centralized sanitation systems. On the other hand, from the 1960s onward, the practice experienced a form of resurgence, driven by a social movement developing at the fringes of, and in opposition to, industrial agriculture.

The End of the Outdoor Toilet?

After the introduction of centralized sanitation systems in cities, the agricultural use of human excreta persisted discreetly in rural areas, following longstanding traditional practices. The primary reason is that running water and sewage networks arrived much later in rural areas than in urban centers. The predominance of flush toilet systems, either connected to sewer systems or managed locally via septic tanks, did not fully take place until the 1970s and 1980s. By the late 1950s, only 54% of residents in urban municipalities with more than 2000 inhabitants were connected to a sanitation network (Bellanger & Pineau, 2010); and by 1960 in the Île-de-France region, only 40% of the sewers and treatment plants envisioned in the general sanitation program for the Paris region had been completed (*ibid.*). It took another 30 years for sewer systems and treatment facilities to become nearly universal (Eliès & IAURIF, 1992). Consequently, cesspools and septic tanks remained in use in rural areas where sewer infrastructure was slow to develop; the same goes for companies responsible for waste removal and its agricultural recycling (see Fig. 11.1). Within this system, excreta are diluted in wastewater, and the

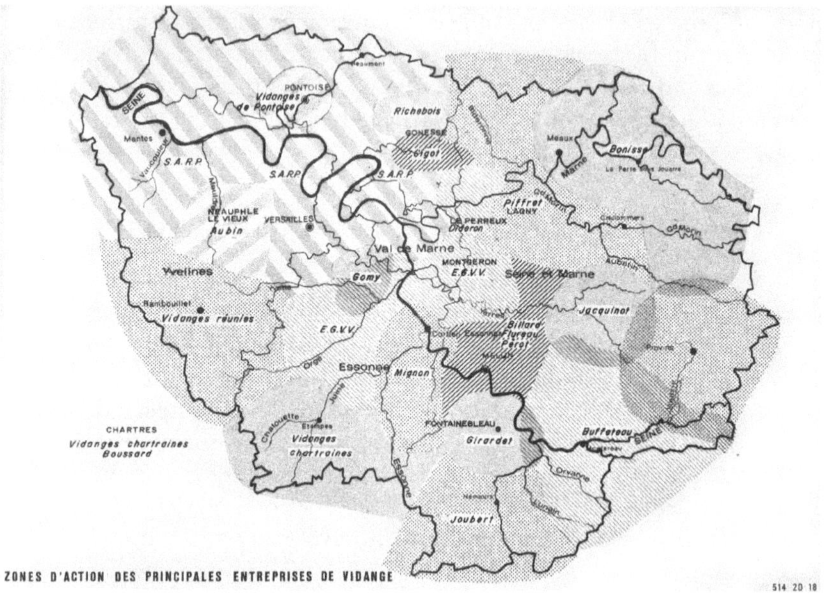

Fig. 11.1 The operational areas of the main waste disposal companies operating in the Paris region during the 1960s. *Source* Financial Agency of the Seine-Normandy Basin, "Disposal of waste products in the Paris region," June (1969), AD92: 2916W478

fertilizing resources they contain are only partially returned to cultivated soils through the remaining channels for the agricultural valorization of sewage sludge. The watercourses that receive treated wastewater struggle to dilute the remaining nutrients after treatment, contributing to their eutrophication (Esculier & Barles, 2019). While the efficiency of wastewater treatment processes has greatly increased over the following decades, the nutrient recycling rate remains low, not exceeding 10% for nitrogen at the national level today (Starck et al., 2024).

The Legacy of Ecological and Eco-Anarchist Movements of the 1970s

During this period, when flush toilets gradually infiltrated rural areas following the introduction of electricity and telephone, alternative voices emerged among architects, landscapers, or agronomists advocating for the return of human excreta to the soil.

While scattered traces of these voices can be found as early as the 1920s, the roots of the contemporary practice of composting toilets are to be found in the 1960s and 1970s within European and American countercultural movements. It is at this moment that recycling human waste emerges as a preoccupation in avant-garde architecture. The reconsideration of organic waste becomes a motif of a social and political critique that unfolds in the squat movement in England and alternative communities in the southwest of the USA, both of which inspired experiments in France. A generation of intellectuals, particularly architects, argued for self-sufficiency and autonomy from networks as a necessary condition for ecological and libertarian ways of life and action (Kallipoliti, 2012; Castillo, 2015). Composting toilets took their place alongside methane digesters, wind turbines, and other techniques for domestic withdrawal from infrastructural networks. They feature in several publications, such as Stewart Brand's "Whole Earth Catalog"[3] from 1968 to 1972, a periodical with a wide audience within countercultural movements and beyond, that combined themes of access to tools, autonomous habitat, and ecological approaches to technology (Kirk, 2007).

On the ground, composting toilets became integral parts of self-construction and autonomous habitat experiments, pioneered in the back-to-the-land initiatives of the 1960s. According to Castillo (2015), the hippie movement developed a distinctive culture of construction that combined a new "materiality of waste" with an individual and collective relationship to work, defined as a "transformative experience open to all." Within this context, these informal autonomous construction practices also fostered new, *metabolic* relationships between the domestic sphere and its environment.

Within this countercultural movement, several standout innovations were widely publicized in the 1970s: the Autonomous House designed by Brenda and Robert Vale in 1975, the Integral Urban House built by the Farallones Institute and architect Sim van der Ryn in Berkeley shortly before 1980, and, before them, Graham Caine's Eco-house in the Woolwich neighborhood of London (1972). While these experiments are

[3] The *Whole Earth Catalog* is a periodical aimed at compiling a comprehensive array of references and addresses on the ecological approach to technology.

recognized in the literature as technical solutions to the surge of ecological concerns following the energy crisis triggered by the 1973 oil shock, they also reflect philosophical positions that articulate a complex fusion of environmentalism with technological appropriation and autonomy (Kirk, 2007).

These experiences resonated in France around the same time, through the development of autonomous architecture using "soft" or "appropriate" technologies. Moreover, in Francophone ecological magazines of the 1970s, these Anglo-Saxon experiments were publicized and explicitly linked to the protection of natural biogeochemical cycles, serving as a critique of the sewage industry[4] (see Fig. 11.2). Some publications stand out, largely influenced by the foreign publications they frequently reference: the work of architect Robert Chareyre (*La maison autonome*, 1980) and that of Pierre Le Chapellier (1979) and the Study Group for the Ecological House. Both authors promoted the reintegration of habitats into ecological cycles via the recycling of human excreta within local agriculture, particularly through composting. One project that materialized was the "Durand farm" in Vachères, Beauce, designed as a bioclimatic architecture experiment. Here, David Roditi installed "mobile" toilets built on rails to facilitate the production of artificial manure (Roditi, 2019).

To conclude, it is worth noting that during this period, as political ecology began making strides on the libertarian front, there was also a pointed critique of the industrial agricultural model. Murray Bookchin, often referenced in the publications mentioned above, devotes an entire chapter to the question of soil fertility depletion in his seminal book "Our Synthetic Environment" (Bookchin, 1962). However, these experiments do not always confront agricultural issues directly; they often reduce the question of nutrition to the scale of the domestic sphere and, in doing so, sidestep the political (and technical) struggle over societal food systems. This is notably the criticism made by environmental and technology historian Joel Tarr, as a contemporary observer in the US context, regarding their emergence (Tarr, 1979).

[4] Examples include: Non-tox. (1975). Le compostage des déchets ou la jouissance dans l'ordure. *La Gueule ouverte*, (43), 6–7; Simonnet, D. (1976). Inventaire des maisons écologiques. *Le Sauvage*, (34), 48–67; Samuel, P. (1972). Pollution et anti-pollution. *Survivre et vivre*. (12), 31–33.

Fig. 11.2 Front cover of the magazine "LE SAUVAGE," Autumn 1977, announcing a special issue on human excrement

The Impossible Generalization of the Sewer System

Beyond these architectural experiments at the domestic scale, from the mid-1970s onward, technical guides related to ecological sanitation began to proliferate, albeit in a slightly different context. They were produced by engineers and architects addressing the issue of sanitation access in poor tropical countries, where the widespread adoption of sewage systems and flush-based sanitation techniques is considered unfeasible. More specifically, their critique focused on the continued exportation of a model they considered flawed: "Despite the known disadvantage of centralized sewerage systems in the industrial nations, they are still being recommended for adoption in cities of the developing world" (Leich, 1980, p. 56).

Technical guides that emerged in the years that followed include "Goodbye to the flush toilet" (Stoner, 1977) and "Stop the 5-gallon Flush" (MCHG, 1976). The introduction of the latter starts with the observation of the technical, environmental, and economic limitations of centralized sanitation based on the sewer system: pollution from the discharge of inadequately treated wastewater into aquatic environments, excessive water consumption, and the economic limitations of network systems, especially in dispersed housing. The impossibility of generalizing the system at the international level, due to a lack of access to water and economic resources to build and maintain such infrastructure, is also highlighted. Managing excreta without water is then presented as a technically and economically viable alternative to the unsustainability of the centralized sanitation system. The move toward a "sewerless society" begins to be seen as a significant step toward a more sustainable future (Leich, 1975).

In France in the 1980s, these technical guides were widely disseminated, including Antoine Laborde's "Practical Manual of Water" (Laborde & Motec, 1980) and Béatrice Trélaün's "Toilets without water, an alternative to the sewer system" (1983). These guides introduced dry toilets to audiences beyond the experimental roots of the movement, sparking a dynamic that continues to grow, where the search for alternatives to the flush toilet-based system, the sewer, and the sewage treatment plant has begun to scale up in recent years.

It was on these foundations that the Réseau de l'Assainissement Ecologique (Network for Ecological Sanitation) was created in 2006, marking an important milestone in the development and diffusion of dry toilets in France. Over time, this network has brought together a diverse group of actors that today constitute a community of practice: manufacturers, installers, rental companies, users, consultants, and environmental educators. Since its inception, it has focused on developing recommendations for professionals and individuals, lobbying policymakers, and facilitating the peer-to-peer exchange of information. The culture of the network has become increasingly professionalized, transforming into an economic sector in its own right. However, it remains an institution that is difficult to categorize. Bringing together businesses, cooperatives, individuals, and several researchers, the network could be considered an example of what Clark and Szerszynski (2021) describe as an "earthly multitude": a community sharing practices that engage with a specific terrestrial stratum (here, fertile soil) and, in so doing, transform it.

11.4 Compost Toilets and Contemporary Domestic Practices Related to Human Manure

Compost toilets have been developing in France for about 40 years. As a viable option for equipping homes, they are now part of a growing movement toward environmental sobriety at the household level. Despite its gradual growth in various contexts, the practice remains mostly domestic or communal and, in terms of numbers, extremely marginal compared with flush toilet-based sanitation systems.

In what context does the management of materials from domestic dry toilets, which mainly involves composting, take place today? Are those who practice it motivated by a desire to overturn the conventional use of fertilizer in industrial agriculture? The question is worth exploring, as the installation of domestic dry toilets does not necessarily imply the intention to use the resulting materials to fertilize a vegetable garden, at least initially.

This final part of the chapter is based on a sociological survey conducted in the Limousin region of France, around the village of Faux-la-Montagne (450 inhabitants). This highland area, dedicated to livestock farming and forestry, has experienced a massive rural exodus during the twentieth century but is now being repopulated with the arrival of new residents. For some, living in rural areas and adopting a lifestyle focused on ecological sobriety constitutes a form of political commitment (Schlosberg & Craven, 2019). In the case of Faux-la-Montagne, the municipality also demonstrates a strong commitment to ecological issues. For example, the temporary housing created for new residents by the municipality is equipped with compost toilets.

Twenty users of dry toilets (men and women) were interviewed over the course of 14 semi-structured interviews (some users living together).[5] The individuals interviewed share a significant cultural background but

[5] The interviews conducted by Mathilde Soyer covered the following topics: background, motivations for installing dry toilets; hygiene, cleanliness, privacy; management organization; relationships with others regarding toilets. They were recorded, transcribed, and subjected to qualitative analysis.

often have modest incomes. A diversity of profiles was sought, based on criteria such as housing configuration, toilet types, and types of material valorization. Interviews were also conducted with professional actors in the sector.

(Re)Discovering the Living Nature of Soil

The individuals interviewed all fall into the category of residents commonly referred to locally as "neo-rurals," a term they also use to describe themselves. They are a non-native population, relatively young, well-educated, and formerly urban residents. Agricultural and food-related issues are at the forefront of their concerns. For most of them, however, the use of the toilet for producing compost was not initially considered. As Marie, a local ecological sanitation professional, explains:

> People initially think: 'What size bucket do I need? ... What do I have to do with it? ... Where am I going to find sawdust?' They often don't think much further than that! Most don't think about the compost bin, yet it's the first thing to do.

Composting dry toilet materials is a gradual process: The correct protocols and recipes are refined over time, based on observation, discussions, and information found on the internet. Users of dry toilets discover that producing human manure from excreta involves facilitating a biological process. They become attentive to properly nourishing their plants and to ensure decomposers produce good-quality compost, they also need to be well-nourished. It's about creating the right environment for the organisms that will perform this transformation, establishing the optimal conditions for their "work": "*Worms need air to live, to do their work. So you have to create an environment for them, you have to make them a habitat. So they can come and go, and go to work.*" (Thierry)

In such a context, what does the term "work" refer to when it concerns worms and other decomposers? What type of relationship does it imply? According to one resident, who is a builder, carpenter, and handyman, it could be understood in the same way as when we say "the earth works" or "the wood works"; that is, in the sense of transformative activity, a vitality of natural elements. For her part, Jacqueline Milliet (2015) proposes an interpretation based on the ambivalence, between friendship and exploitation, of this working relationship with decomposers, based on her study of users of apartment vermicomposters. This relationship generally refers to a form of companionship that also involves identification with the beings wriggling in the bins, these "compost artisans" such as worms, bacteria, mites, and fungi, as they are referred to by compost expert Eric Sabot (2005). According to Joseph Jenkins, author of a reference guide on dry toilets, composting is likened to "feeding" visible and invisible organisms, with compost viewed as the product of their digestion (Jenkins, 1994).

Moreover, composting is an activity that sparks wonder and allows for creative expression, where one is producing tsingnot bread or beer but soil (using expressions like "*my own vintage*" of soil).

> I have a knack for it (composting). I don't know why. I love getting my hands dirty, mixing, stirring... and I'm convinced it works. When you put yourself into it, it grows better; it's obvious, I think it is obvious. It rebalances itself. And when I do that, there's a connection to life because I imagine all the bacteria, I see the critters, and I connect to the idea of the cycle. And I'm happy when I do it when I have my hands in it. (Léo)

Observing and accompanying the composting process gradually demystify the decomposition of excrement. The realization that materials from human bodies can transform into "soil" is accompanied by a process of familiarization, fostering a growing interest in the living beings of the soil, with whom a sustaining connection is established.

A Dimension of Autonomous Housing

Once the pile of compost begins to accumulate, the question of what to do with it becomes pressing. This is when the possibility of using it as manure makes itself felt. Not everyone chooses to fertilize edible plants in the vegetable; a good portion of respondents opt to spread it at the base of hedges, ornamental plants, or even fruit trees. The main motivation behind this circuitous route back to the soil, sometimes temporary while mastering the process, is due to a lack of confidence in its safety. The cycle in which the practice of dry toilets is embedded extends beyond a simple closed loop between the human body and the vegetable garden. The connection to the soil and its fertility seems broader than a strictly productivist concern with growing bigger pumpkins. Thierry underlines the political dimension he associates with managing (his own) waste:

> When I go to empty the bucket, I say that I'm performing a 'crap action' [merdic'action]. It's a moment, a gesture, where you manage your waste but you're also doing something that will serve the living. There's a connection to the cycle. You're connected.

The life of the composted materials, once integrated into the soil, surprises them, and what they marvel at, ultimately, is what doesn't depend on them: "*I scatter the compost in my garden, and I have tomatoes growing everywhere, even though I've never planted any. And I eat them.*" (Thierry).

In conclusion, what the inhabitants of Faux-la-Montagne share through the practice of *managing their toilets* is a pursuit of a certain autonomy at the domestic scale, reconfiguring their relationship with the land and its fertility. Off-grid housing is not about creating a bubble isolated from the world but a particular way of inhabiting a place, a milieu, and connecting to its natural cycles (water, soil, etc.). The notion of autonomy in contemporary practices of dry toilets is not about maintaining closed loops for the sake of efficiency. Instead, it is a process that involves cultivating attention for one's habitat and the surrounding environment, in the dual sense of "being attentive to" (observing and learning) and "taking care of" (protecting). This practice becomes a way to rejuvenate the land by revitalizing the soil and everything that grows from it.

11.5 Conclusion

At the end of our exploration, it appears that in France, contemporary domestic practices associated with composting toilets diverge from past logics of circular management of urine and feces. The nineteenth century and World War II marked two moments of explicit promotion of human excrement as fertilizing materials. During these times, human excreta were integrated into the fertilizer market like other resources and reused by farmers on their farms, with their collection and circulation supported by the political, scientific, and technical institutions of the day.

After these large-scale agricultural uses of human manure had been abandoned, the reconsideration and revalorization of human excreta emerged within the sphere of experimental approaches to autonomous housing, operating at a completely different scale. Since their emergence in the 1960s within various back-to-the-land movements and the urban squat movement, the use of individual composting toilets has continued to expand on the fringes, gradually becoming normalized to the point that, in 2009, they became an officially recognized, legal sanitation system. Initially developed mainly in rural, domestic contexts, composting toilets are now found in much more diverse settings, serving an increasingly varied user base. For example, in urban areas, composting toilets have been integrated into cooperative housing projects (Joveniaux et al., 2021), and their use has expanded beyond the domestic sphere to include festivals, concerts, public toilets, restaurants, and schools. Both private and public actors are working to establish composting toilets as a structured industry sector. This includes pilot-building projects, experimentation, and planning at the local and regional levels. Industrial interests are also positioning themselves to develop treatment processes, particularly for urine. However, in practice, most of these initiatives do not result in a functional supply chain for human manure; it tends to be only the local connection between dry toilet hire companies for events and farmers that has so far led to human manure being used in the field. In this context, in a rich country like France, where agriculture continues to rely on an industrial model dependent on an abundance of synthetic and extracted fertilizers, the use of human manure remains largely confined to the margins of both sanitation and food production systems.

For now, the practice of composting toilets is more often associated with experiments in domestic autonomy and sustainability that integrate daily life into the cycles of living beings. It is in the context of habitat and home that critical perspectives concerning soil fertility are explored. The soil, as a living environment, becomes part of a broader commitment linked to the land and its inhabitation, in which the practices observed are part of a wider "sustainable materialism" movement (Schlosberg & Craven, 2019). The subversive potential of using human manure in relation to the current agricultural model can be seen as a challenge to the extractivist model of fertilization. The metabolic relationships between town and country have essentially broken down, and the economic model that supported these relationships has disappeared. Starting from the margins—in this case, autonomous housing—with all the limitations of its marginality, it could be argued that the lever for transformation is to be found not in the large-scale substitution of one type of fertilizer by another in an existing market but rather in the proliferation of alternative forms of housing, community, and agricultural production that explore less damaging relationships with the living environment. Different approaches to economic exchange are also emerging, anchored in their material and, more specifically, biological ecologies (Moore, 2015).

Anna Tsing (2012) uses the term "scalable" to describe production systems that can be scaled up and scaled again, without transforming themselves or having to rethink their basic elements. As far as agricultural production is concerned, this often entails a complete negation of the wider cultural and biological diversity of place—an indifference so central to the contemporary world that Tsing calls Plantationocene. Here, by contrast, the new forms of relation being developed—to the soil and the entities and phenomena involved in renewing its fertility—retain their singularity in every place in which they are deployed. The question of what the soil is, and what the work done by the beings who live in it might be, remains open. The very materiality of ***human shit*** is key here: It blurs the boundaries between inside and outside, between public and private, human and non-human, making it a potential transformer not only of our ideas of habitat but also, more broadly, of ways of inhabiting the world, of making place. The question of production remains open. It's a question that is answered, in practice, in different ways, in different places.

Finally, while the question of large-scale substitution of mineral and synthetic fertilizers by human excreta is not yet at the center of public debate, its day might soon arrive. The fragilization of the agri-food system that is currently happening is quite profound; in addition to the sharp rise in the costs of fertilizer and fuel, there are the growing effects of climate change in regions that have highly specialized agricultural production, leading to increased risks of bad harvests and food shortages. At this juncture, the crises facing the global and industrial agri-food system will surely lead to it being called into question and inciting a radical rethink of how human excreta is managed at different scales (domestic and territorial). At the domestic scale, the practice of dry toilets has a critical potential concerning the question of soil fertility and the integration of the human habitat with its environment.

But is there any way of scaling up the agricultural use of human manure other than by re-industrializing it? There are no easy answers here. The initiatives that are currently emerging reflect sometimes conflicting visions of the world and are based on economic, political, and organizational approaches that remain quite diverse. Everything will depend on the status accorded to excremental matter and the types of relationships that develop with and around it.

Competing Interests and Acknowledgements This work was carried out with the financial support of the Seine-Normandie Water Agency (AESN) and the Paris Municipality.

References

Abdon, A. (1948). *Dispositif séparateur de matières excrémentielles.* (Patent No. 940.350). Filed 5 February 1947; issued December 9, 1948.

Adler, E. (2020). *La gestion des déjections humaines: un défi urbain. Le cas de la ville de Lyon, de la fin du 18e au début du 20e siècle.* (Doctoral dissertation). Université Paris-Est.

Financial Agency of the Seine-Normandy Basin (Agence Financière de Bassin Seine-Normandie). (1969). *Élimination des produits de vidange dans la région parisienne.* AD92, 2916W478.

Aguerre, J.-P. (2003). Scatophobie des villes, scatophilie des champs. Gestion et utilisation des fèces à Lyon à la fin du 19e siècle. *Ruralia: Sciences sociales et mondes ruraux contemporains*, 12/13. http://journals.openedition.org/ruralia/329

Barles, S. (2005a). Experts contre experts: les champs d'épandage de la ville de Paris dans les années 1870. *Histoire urbaine, 14*(3), 65–80.

Barles, S. (2005b). *L'invention des déchets urbains: France, 1790–1970.* Champ Vallon.

Barles, S. (2021). Les excreta urbains: entre valorisation et abandon (fin XVIIIe-début XXIe siècles). *Raison présente, 220*(4), 9–18.

Barles, S. (2022). De la chimie urbaine à l'écologie territoriale: deux siècles et demi d'analyse du métabolisme urbain. In C. Granjou & J. Salomon Cavin (Eds.), *Quand l'écologie s'urbanise* (pp. 63–94). UGA Éditions.

Bauer, C. (1942). *La fertilité par les engrais organiques.* Guy Le Prat.

Bellanger, E., & Pineau, E. (2010). *Assainir l'agglomération parisienne: histoire d'une politique publique interdépartementale de l'assainissement (XIXe-XXe siècles).* L'atelier.

Bernard, L. (2019). La médecine néo-hippocratique des années 1930: le temps d'une rencontre. *Histoire, médecine et santé, 14*(3), 63–81.

Bookchin, M. (1962). *Our synthetic environment.*Alfred A. Knopf. (originally published under the name of L. Herber).

Boyer, J. (1945). Le fumier artificiel, sa fabrication en usine. *La Nature, 3081*(février), 44–46.

Burgevin, H. (1940). Le fumier artificiel. *Revue des Agriculteurs de France, 72*(12), 228–229.

Caine, G. (1972). A revolutionary structure. *OZ magazine, 45*, 12–13.dussea

Carnino, G. (2013). L'environnement et la science. Acclimater la population de Gennevilliers aux débordements des eaux usées parisiennes, 1870–1880. In M. Letté & T. Le Roux (Eds.), *Débordements industriels: Environnement, territoire et conflit (XVIIIe-XXIe siècle)* (pp. 199–223). Presses universitaires de Rennes.

Castillo, G. (2015). Hippie modernism: How Bay Area design radicals tried to save the planet. *Places Journal*. https://placesjournal.org/article/hippie-modernism

Cépède, M. (1961). *Agriculture et alimentation en France durant la IIe Guerre Mondiale*. Editions M.-Th. Génin.

Chareyre, R. (1980). *La Maison autonome*. Éditions Alternatives.

Chouard, P. (Ed.). (1942). *Le Jardin de la famille. Légumes, arbres fruitiers, petits élevages*. Éditions du Secours National.

Clark, N., & Szerszynski, B. (2021). Planetary multiplicity, earthly multitudes: Interscalar practices for a volatile planet. In G. Dürbeck & P. Hüpkes (Eds.), *Narratives of scale in the Anthropocene* (pp. 75–93). Routledge.

Coutard, O. (2010). Services urbains: la fin des grands réseaux? In O. Coutard & J.-P. Lévy (Eds.), *Écologies urbaines* (pp. 102–129). Economica.

Demolon, A., & Burgevin, H. (1937). La fabrication du fumier artificiel. *Bulletin de la Société d'Encouragement pour l'Industrie Nationale, 136*(7–8), 297–300.

Desjobert, G. (1950). Une méthode de fabrication du fumier artificiel. *Bulletin technique d'information des Ingénieurs des Services Agricoles, 46*, 23–26.

Dufour, E. (2024). *La fin du recyclage? Rupture métabolique et politiques biogéochimiques en région parisienne au XXe siècle*. (Doctoral dissertation). Université Paris-1 Panthéon-Sorbonne.

Dufour, E. (2023). Entre le regain et l'incendie: l'étape oubliée du compostage industriel, voie médiane abandonnée du traitement des ordures ménagères (Île-de-France, années 1940 – années 1990). *Flux, 131*(1), 32–50.

Dumont, R. (1946). *Le problème agricole français. Esquisse d'un plan d'orientation et d'équipement*. Les Éditions Nouvelles.

Dusseaulx, R. (1943). Comment remplacer les engrais chimiques? In J. Le Roy Ladurie (Ed.), *La Terre. Encyclopédie Paysanne*. Flammarion.

Eliès, A., & Institut d'aménagement et d'urbanisme de la région d'Ile-de-France. (1992). *Assainissement et dépollution des eaux en Ile-de-France. Production et gestion des résidus (boues, graisses, sables, matières de vidanges...)*. (Master's thesis). Ecole Nationale Supérieure Agronomique de Rennes.

Esculier, F. (2018). *Le système alimentation/excrétion des territoires urbains: régimes et transitions socio-écologiques*. (Doctoral dissertation). Université Paris-Est.

Esculier, F., & Barles, S. (2019). Past and future trajectories of human excreta management systems: Paris in the nineteenth to twenty-first centuries. In N.

Flipo, P. Labadie, & L. Lestel. (Eds.), *The Seine River Basin* (pp. 117–140). Springer.

Foster, J. B. (1999). Marx's theory of metabolic rift: Classical foundations for environmental sociology. *American Journal of Sociology, 105*(2), 366–405.

Gorman, H. S. (2013). *The story of N: A social history of the nitrogen cycle and the challenge of sustainability*. Rutgers University Press.

Hawkins, G. (2004). Shit in public. *Australian Humanities Review*, 31–32. http://australianhumanitiesreview.org/2004/04/01/shit-in-public

Jenkins, J. (1994). *The humanure handbook. A guide to composting human manure*. Jenkins Publishing.

Joveniaux, A., de Gouvello, B., & Legrand, M. (2021). L'émergence d'un commun en matière d'assainissement urbain: les toilettes séparatives en habitat participatif. *Flux, 124–125*, 27–40.

Kallipoliti, L. (2012). From shit to food: Graham Caine's eco-house in South London, 1972–1975. *Buildings and Landscapes, 19*(1), 87–106.

Kawa, N. C., Ding, Y., Kingsbury, J., Goldberg, K., Lipschitz, F., Scherer, M., & Bonkiye, F. (2019). Night soil. Origins, discontinuities, and opportunities for bridging the metabolic rift. *Ethnobiology Letters, 10*(1), 40–49.

Keilling, J., & Casays, L. (1953). Le problème du traitement et de l'utilisation des eaux usées. In Commissariat Général au Plan (Ed.), *Deuxième plan de modernisation et d'équipement. Rapport des Commissions de la production agricole et de l'équipement rural, october 1953*(p. 17), annex III.

Kirk, A. (2007). *Counterculture green: The whole earth catalog and American environmentalism*. University of Kansas Press.

Kohlbrenner, A. (2014). De l'engrais au déchet, des campagnes à la rivière : une histoire de Bruxelles et de ses excréments. *Brussels Studies*, *General collection*, *78*. https://doi.org/10.4000/brussels.1224

Laborde, A., & Motec, J. (1980). *Manuel pratique de l'eau: promenade en eau trouble*. Éditions Alternative et Parallèles.

Lacroix, P. (1943). La recrudescende actuelle de l'ascaridiose. *Le Concours Médical, 65*(31), 609.

Le Chapellier, P. (1979). L'eau, le recyclage et l'énergie. *L'Affranchi*, *69*.

Legrand, M. (2020). Digestions fertiles? Le retour au sol des excréments humains. *Revue d'Anthropologie des connaissances*, *14*(4). https://doi.org/10.4000/rac.11042

Legrand, M., Joveniaux, A., Arbarotti, A., de Gouvello, B., Esculier, F., & Tabuchi, J. P. (2021). Séparation à la source et valorisation des excrétats humains du Grand Paris: des filières émergentes. *Techniques, Sciences et Méthodes, 2021*(9), 103–118.

Leich, H. H. (1975). The sewerless society. *Bulletin of the Atomic Scientists, 31*(9), 38–44.

Leich, H. H. (1980). Appropriate sanitation for human settlements in Africa. In *6th WEDC conference. Water and waste engineering in Africa.*

Le Noë, J., Billen, G., Esculier, F., & Garnier, J. (2018). Long-term socio-ecological trajectories of agro-food systems revealed by N and P flows: The case of French regions from 1852 to 2014. *Agriculture, Ecosystems & Environment, 265*, 132–143.

Maïs, H. (1946). Peut-on maintenir la fertilité des terres sans engrais chimiques? *La Science et la Vie, 342*, 125–130.

Martin, P. (2017). Twentieth-century fertilizers in France from natural mixing to artificial making (1890–1970). *Cahiers François Viète, III*(2), 69–91.

Milliet, J. (2015). Le lombricomposteur d'appartement, les déchets et la terre urbaine. Le cas de la Suisse romande. *Revue d'ethnoécologie, 8*. https://doi.org/10.4000/ethnoecologie.2305

Minimum Cost Housing Group. (1976). *Stop the five gallon flush. A survey of alternative waste disposal systems*. McGill University.

Moore, J. W. (2015). *Capitalism in the web of life: Ecology and the accumulation of capital*. Verso Books.

Pessis, C. (2020). Histoire des "sols vivants": Genèse, projets et oublis d'une catégorie actuelle. *Revue d'anthropologie des connaissances, 14*(4). https://doi.org/10.4000/rac.12437

Puget, R. (1939). Evacuation et désintégration des matières fécales et ordures ménagères - Evacuation des eaux usées par des moyens de fortune ou des procédés pratiques et peu couteux. *L'hygiène sociale. Revue médicale internationale, 10*, 275.

Roditi, D. (2019). *The life of an unknown artist*. p. 180. http://davidroditi.com/unknownartist.pdf. Accessed November 29, 2023.

Sabot, E. (2005). *La pratique du compost et des toilettes sèches*. Maison autonome.

Saitō, K. (2017). *Karl Marx's ecosocialism: Capitalism, nature, and the unfinished critique of political economy*. Monthly Review Press.

Schlosberg, D., & Craven, L. (2019). *Sustainable materialism: Environmental movements and the politics of everyday life*. Oxford University Press.

Smil, V. (2001). *Enriching the earth: Fritz Haber, Carl Bosch, and the transformation of world food production*. MIT Press.

Starck, T., Fardet, T., & Esculier, F. (2024). Fate of nitrogen in French human excreta: Current waste and agronomic opportunities for the future. *Science of the Total Environment, 912*, Article 168978.

Steffen, W., Richardson, K., Rockström, J., Cornell, S. E., Fetzer, I., Bennett, E. M., Biggs, R., Carpenter, S. R., Vries, W. de, Wit, C. A. de, Folke, C., Gerten, D., Heinke, J., Mace, G. M., Persson, L. M., Ramanathan, V., Reyers, B., & Sörlin, S. (2015). Planetary boundaries: Guiding human development on a changing planet. *Science* 347, 1259855.

Stoner, C. H. (1977). *Goodbye to the flush toilet Water-saving alternatives to cesspools, septic tanks, and sewers*. Rodale Press.

Tarr, J. A., & Dupuy, G., (1988). *Technology and the rise of the networked city in Europe and America*. Temple University Press.

Tarr, J. A. (1979). Review of Sim van der Ryn, The toilet papers: Designs to recycle human waste and water: Dry toilets, greywater systems, and Urban Sewage. *Technology and Culture, 20*(1), 231–333.

Tomic, S. (2017). La "science des engrais" et le monde agricole en France au dix-neuvième siècle. *Journal for the History of Environment and Society, 2*, 63–93.

Trélaün, B. (1983). *Water sans eau : Alternative au tout à l'égout*. Éditions Alternatives.

Tsing, A. L. (2012). On nonscalability: The living world is not amenable to precision-nested scales. *Common Knowledge, 18*(3), 505–524.

CHAPTER 12

Redesigning the Cause: From Food Waste to Compost in Entrepreneurial Collection and Composting Businesses

Maud Hetzel

12.1 Introduction

Researchers have shown a growing interest for soils in the past 10 years and have increasingly considered them as an environmental "critical issue" (Meulemans & Granjou, 2020). The research on soils has also led to redefine them: they are now acknowledged as living beings and no longer as inert matter thanks to the soils' life "rediscovery" (Puig de la Bellacasa, 2014). Meanwhile, some empirical studies have explored new practices on organic matter, such as waste–derived fertilizers (Lupton, 2017), food surplus (O'Brien, 2012), grape marc and vine wood reuse (Krzywoszynska, 2012), or human *excreta* (on this matter, see Chap. 11 by Legrand et al.). My doctoral work looks into emerging composting activities: more specifically, I have investigated entrepreneurial businesses,

M. Hetzel (✉)
Centre Georg Simmel, EHESS, Paris, France
e-mail: maud.hetzel@ehess.fr

© The Author(s), under exclusive license to Springer Nature Switzerland AG 2025

O. Labussière et al. (eds.), *Back to the Ground*,
https://doi.org/10.1007/978-3-031-88888-5_12

whose purpose is to collect and manage organic waste—mostly from food professionals and retailers.

From the point of view of the actors on the field, if you are concerned with environmental issues—especially those related to food and waste management—you are necessarily anxious to protect soils and to favor their organic fertilization. In this perspective, they describe food-waste composting as a return to simplicity, in contrast to the absurdity and nonsense of waste business, and especially incineration. However, soil impoverishment issues and soil organic fertilization are not some well-known and newsworthy "public problem" (Gusfield, 1997). Unlike many ecological issues, "there are no binding conservation measures in place for soils that would help preserve their proper functioning and the vast diversity of living organisms, including microscopic ones, that inhabit and contribute to their formation"[1] (Fournil et al., 2018). Indeed, organic waste collection and management companies were established in response to regulatory obligations regarding food waste, implemented in a broader context of "throwaway societ[ies]" (Evans, 2012), where the abundance and nature of waste have become "unmanageable" and "uncontrollable" (Monsaingeon, 2017, p. 82). By harnessing organic waste sources, which are both potential drivers for more environmentally friendly waste management and a potential resource for emerging economic sectors, business creators are waste composting professionals more than soil preservation and organic fertilization activists. As Marine Legrand et al. demonstrate in the chapter dedicated to human manure (see Chap. 11) the centrality of waste in relation to soil fertilization is the result of historical developments. Previously required due to fertilizer shortages, composting of organic materials is now justified by the overproduction of waste.

How to understand the discrepancy between the issues of soil preservation lying in the background while food-waste composting practices are in the foreground? Despite this gap, how can business creators seize the soil conservation issue, and therefore challenge the prevalence of waste management over issues related to soil?

I will address these questions by considering waste and soils as "causes," which allows for the integration of different levels of analysis. Speaking in terms of "causes" first enables to mobilize the concept of

[1] My translation.

"cause field." Coined by Bereni (2021), this notion invites to focus on the logic behind the construction of a cause and the nature of the actors who take it up. I argue that this scientific concept can be articulated with that of "cause entrepreneurs." This concept stems from resource-mobilization approach (McCarthy & Zald, 1977): it allows for an analysis at the individual level by paying attention to personal commitment motives. Finally, I also consider the cause in action, which means examining the various ways in which companies concretely implement this cause. Combining these different scientific approaches helps to think both the cause's institutional structure, its organizational reality and the individual commitment to the cause in the same movement, while other sociological studies tend to mobilize either of these approaches.

The analysis presented here is based on an ethnographic inquiry conducted within two companies, Bikompost and Magicompost, both of which are located in several major French cities. While Bikompost is considered one of the pioneers in bicycle waste collection, Magicompost is one of the most developed companies in the sector. Their activities are similar—they compost food waste from businesses and households with the aim of ecological waste management—but they have different economic and technical practices. Observations were carried out in these two companies through regular voluntary participation in waste collection activities at Bikompost, from November 2019 to March 2020, and by becoming part of the working collective at Magicompost, particularly on the administrative side, from June to December 2020. These two periods of intensive observation, involving participation in various daily tasks (meetings, administrative and logistical work, and waste collection) were complemented by interviews that focused on the professional background of the people, their daily work in the organization, and their relationship with environmental movement. A total of 61 interviews were conducted in the two companies, 43 at Magicompost and 11 at Bikompost.

In this chapter, I show how the soil conservation cause and the food-waste composting cause compete and complete each other. In a first section, I will argue that the cause field is more structurally organized around waste than around soils (Sect. 12.2). Despite this structural prevalence of waste, I will, in a second section, suggest that the more business creators practice composting, the more they learn and become aware of the soil conservation issue (Sect. 12.3). Finally, I will highlight that the two companies offer two heterogenous—and unequally legitimated—interpretations of the cause (Sect. 12.4).

12.2 A "Circular Economy"? Treating Waste Before Implementing Organic Fertilization

By using the term "circular economy," public institutions avoid addressing the issue of waste, which would systematically transformed into "resources." However, the collection and composting activities primarily revolve around waste, whereas the resulting organic amendment is often surrounded by uncertainties regarding its use and quality.

A Waste-Structured Sector

The businesses I studied can be seen as waste reduction initiatives—much more than organic fertilizer producers. In order to demonstrate this initial discrepancy, I propose to consider the genesis of both composting food-waste sector and businesses.

Firstly, at the institutional level, regulations regarding waste producers encouraged the implementation of composting services. While the literature on "public issues" often portrays legislation and public policies as the ultimate goal of mobilizations (Neveu, 2017), here regulations are seen as a starting point rather than an end in themselves. In this regard, my research resonates with the "battle for gender equality" studied by Bereni (2021), where regulations revive mobilizations rather than putting an end to them.

Indeed, regulatory texts have created new economic opportunities and have served as the main driving force behind the emerging commercial sector for the past 15 years. Source separation and biological valorization have been mandatory since 2010 for "individuals or entities producing or holding significant amounts of waste consisting mostly of biodegradable waste",[2] specifically for producers generating more than 10 tons of waste per year. These sorting obligations have been extended to all producers starting since January 1, 2024. It is only since the implementation of these regulations that a range of commercial collection services has emerged in France. The regulations, which pertain to waste rather than soils, serve as a commercial argument.

Secondly, in terms of individual trajectories, the creators were primarily driven by an interest in waste reduction. Even before starting their businesses, the founders have demonstrated an interest in the issue of waste,

[2] Grenelle 2 Act, number 2010-788, 12 July 2010.

while having limited knowledge about soil-related matters. In this regard, the question of waste is spontaneously mentioned during interviews to explain the motivation behind the business creation. Moreover, some business creators are politically committed toward environmental issues, and more specifically in favor of waste reduction. Creating a business in the waste sector is then aligned with their associative and personal commitments. For example, Elsa, the founder of Bikompost, explained in an interview that at the time of starting her business in 2015, she was "deeply involved in food-related issues, and particularly food waste." In addition to her environmental activism, she also sought an individual solution for composting her household waste. Since her building lacked a composting facility, she would cycle her waste to her parents' place, which was ten kilometers away. The creation of a business focused on the collection and composting of food waste stemmed from this reflection on organic waste and the realization of the challenges faced by households in composting their waste in urban areas:

> So, I started to have this idea like, 'Damn, there are very few composting solutions available for residents.' There are neighborhood composters, community composters, but there aren't many, and it's for people who really want to go there (she laughs) because there are specific hours, and it's only in certain neighborhoods. So, at first... Where I lived, there was no composter, so I would take my bike and bring my food scraps to my parents' place in [the suburbs of the city]. And actually, when talking to people around me, I thought, 'It would be interesting if we could collect waste by bike for the residents'. (Interview with Elsa, founder and manager of Bikompost, 32 years old)

Other founders of businesses mention the economic prospects related to waste in general and consider various opportunities for entrepreneurship, all within the waste sector. For instance, Cédric, the founder of Magicompost, initially considered creating a company that produced insulation materials from recycled paper before deciding to establish a business in the field of food waste. Similarly, Adrien, the founder of Magicompost in another French city, had "two small entrepreneurial ideas" related to waste, without any real activist interest in the subject:

> I had studied two subjects. The first one was the recycling of reclaimed wood in the platform where I worked with [his former company], where we had all our merchandise. I saw a lot of pallets, crates circulating, and

> there was quite poor waste management for green waste, pallets, external crates, etc. It's made of lime wood. And the waste streams were quite interesting to study. So, I had looked into that. [...] And the second entrepreneurial project that I wanted to start was about reusing boat waste. There are boat hulls that are rotting and not being used much, with a very resistant composite. And currently, there is a sector that is organizing the sorting and reusing of boats. (Interview with Adrien, founder of Magicompost, 28 years old)

The awareness of an ecological public problem and a lack of solutions primarily lies on the side of waste. This initial finding is reinforced by the economic structure of the organizations, which derive income from the collection and treatment of waste rather than the sale of the produced resources.

A Waste-Financed Activity

The economic revenues of the businesses also reflect the centrality of waste in comparison to the compost production. The gap between the revenue generated from waste collection—the service paid by waste producers—and the revenue generated from the sale of compost as organic fertilizer confirms this disparity. In 2020, Magicompost generated a revenue of 230.000 euros from waste collection, while compost sales amounted to 41.000 euros, representing a little over 15% of the collection revenue. At Bikompost, this gap is even more pronounced: compost sales revenue amounted to only 900 euros, which is equivalent to 2% of the collection revenue, amounting to 43.000 euros in 2020 (see Fig. 12.1). How to understand the fact that compost has low economic value, while being presented as a high-quality organic resource?

	Bikompost	**Magicompost**
Revenue from waste collection operations (in euros)	43 000	230 000
Revenue from compost sales (in euros)	900	41 000

Fig. 12.1 Revenue distribution by sector (*Sources* 2021 annual reports)

On one hand, the high price of the collection service is considered justified and justifiable to customers by the sales managers, as a local and high-quality service—and as the saying goes, "quality comes at a price" explained to me a sales manager from Magicompost. On the contrary, compost is sold at a low price, or sometimes even given away for free. Moreover, companies can struggle to find buyers.[3] The difficulty in finding buyers can partly be explained by the location of the studied businesses. Since their activity is to collect and compost food waste in urban areas, the processing facilities are primarily situated in urban centers or close to the outskirts, at a great distance from agricultural zones. The sale of compost then faces logistical challenges—how to transport the materials—and financial challenges—who pays for the transportation of these materials. In this geographical context, it seems difficult to establish a true "metabolism" between the city and its rural areas (Barles, 2009). This difference between the price of composting and the price of compost underscores the shift brought about by the industrialization and centralized management of waste. In the nineteenth century, the flow between cities and the countryside revolved around the sale of organic fertilizers (Barles, 2016). Nowadays, it is the collection of waste that finances the activity, rather than the sale of the organic resources produced.

However, it is worth noting the emergence of urban marketing methods for compost, in the form of small packets sold in gardening stores. These small quantities, sold at a high price, constitute the primary source of compost-related revenue at Magicompost. Nevertheless, they have sparked a number of criticisms, outside and within the company itself—I will bring it up again in the last section of this chapter.

[3] The structural predominance of the issue of waste over the issue of soils can be observed in the very process of the investigation, particularly in the difficulties encountered as a researcher: finding information and contacts related to a potential composting industry proved to be challenging. This failure should be seen as a "valuable clue" (Becker & Richards, 2007): the impossible investigation into compost reveals the limited agronomic and economic structure surrounding this product and reinforces the idea of a space organized around the issue of waste.

12.3 Becoming Cause Entrepreneurs thanks to Empirical Practices

Despite the discrepancy between the cause of food waste and the cause of soils, the practice of composting leads to the discovery of ecological issues related to compost and soils. While sociological studies on "public issues" have often highlighted how the historical context fosters the emergence of a cause (Gusfield, 1997; Henry, 2007), here I aim to understand how the individuals encountered learn about soil through the practice of composting, despite a structural framing that favors waste. How do compost practitioners, through their engagement with materials, become entrepreneurs of the soil conservation cause?

Making Compost and Learning About Soils

The actions necessary for composting—sorting organic waste, observing its decomposition—require attention to the transformation of waste into new organic matter. Studies in the humanities and social sciences dedicated to domestic composting have emphasized the effects of engaging in this transformation. Composting one's waste allows for the "domestication of decay processes" that is, acquiring "a series of gestures, learning, skills, and ways of seeing"[4] (Granjou et al., 2020) that go well beyond the recommendations of public authorities. These domestic practices are often analyzed as forms of "play" (*ibid.*) and enjoyment (Lehec, 2018), far removed from technical prescriptions.

The composting processes I have studied differ significantly from household waste management, as they are professional activities regulated by strict sanitary regulations. The technical directors and managers emphasize this point in interviews to underscore and claim the professionalism of their activities: compost must meet standards and be produced on platforms with sanitary approval. Nevertheless, despite the emphasis on technical and regulatory processes in interviews, a long-term investigation allows for a description of empirical learning experiences regarding composting, gradually acquainting individuals with compost as an organic material. Following Maslen (2015, 2016), I refer to this as "sensory work" on organic matter, which requires looking, smelling, and touching

[4] My translation.

to assess its quality and potentially adapt the process through sensory and practical experimentation.

Despite the diversity of technical processes between manual composting and industrial composting, the handling of organic material in different companies involves practical and cumulative learning. Skills in composting are acquired primarily through empirical tests to discover what technical supervisors refer to as the "right recipe." This expression refers to the ratio between nitrogen-rich materials—the food waste—and carbon-rich materials—the compost shredders. Once food waste is collected, the transformation of organic materials into compost is facilitated by the addition of a so-called dry material, called shredders or mulchers, which helps absorb the moisture from the organic waste and aerate the material. While the nitrogen/carbon ratio draws on certain scientific knowledge, it is primarily determined through sensory work conducted by those responsible for composting. Once the material begins to decompose, its daily observation helps determine if the recipe is good or not. Technical supervisors rely on visual and olfactory indications, in particular: a recipe that is not working well results in a "very wet and dense" material, emitting "strong" or even "awful" odors. In such cases, technical supervisors try to adjust the process, including turning the compost again to aerate it and/or adding more shredders. These ongoing adjustments raise awareness among supervisors about the importance of high-quality compost and also make composting a playful and engaging process.

These practical trial and error generate an interest in the organic process itself. They are accompanied by a growing theoretical interest in soil-related matters, leading to reading books and watching documentaries on the subject. Engaging in composting activities and experiencing the process firsthand leads to both empirical and theoretical learning. These learning experiences lead compost practitioners to emphasize, in interviews, the living nature of soils and composts, as well as the beneficial role of certain organisms in the process—bacteria, worms, etc. Convinced by their empirical and theoretical learning, these practitioners become advocates for compost in a context where the agronomic quality of this product is being questioned.

Advocating the Quality of an Unrecognized Product

Compost is an agricultural product that is often not well-recognized, and it is frequently confused with other composts, such as those produced by mechanical–biological treatment (MBT) facilities.[5] Though all composts are regulated by an AFNOR standard,[6] which sets maximum thresholds for heavy metals, plastics, and certain chemical components, there have been controversies surrounding the AFNOR standard among waste industry companies and environmental associations since the 1990s.

Originally implemented in the 1970s, the AFNOR standard for fertilizing materials was based on "characterization criteria without prescribing specific threshold values or treatment processes"[7] (Daniel, 2021), which led to criticisms in the 1990s as it allowed for the spreading of low-quality compost. A new version of the standard was introduced in 2006. Nevertheless, the thresholds defined by the standard still remained "very favorable" (*ibid.*) to industrial composting solutions, partly due to the involvement of waste industry professionals. Consequently, the controversy between industrial companies and environmental associations persisted even after the standard was reformed. Environmental associations implemented various methods to challenge the standard as defined, including conducting competitive evaluations or "sensorial demonstrations" of the allowable levels of metals and plastics in standardized compost (*ibid.*). The alerts raised by environmental associations eventually led to the prohibition of the sale of industrial composts produced by mechanical–biological treatment (TMB) facilities starting in 2027.

This brief overview of the standard highlights that it does not clearly establish a definition of what would be a "good" compost. The compost produced by the studied companies is affected by the negative image associated with industrial composts:

> As a result, these MBT machines completely discredited food compost. It is associated with very poor-quality compost because it is linked to this

[5] MBT composts are generated from non-source-separated household waste and may contain significant levels of plastic and glass.

[6] AFNOR stands for the French Organization for Standardization and represents the French government at the International Organization for Standardization (ISO). Its duty is to implement standards regarding environment, health, security, etc.

[7] All quotations from François-Joseph Daniel's article (2021) have been translated by the author.

> industrial compost. Today, it is common knowledge that this compost is bad... Agronomists know very well that compost from MBT is standardized... So, they are aware that the standard is not worth much. (Interview with Cathy, founder of Bikompost, 56 years old)

While economic sociology works have shown that products are rarely presented "bare" to customers and that they are "framed by multiple metrologies"[8] such as instructions and certifications (Vatin et al., 2013, pp. 22–23), the NFU 44-051 standard does not serve as a reliable point of reference regarding compost quality. On the contrary, compost standardization highlights that metrologies can "produce more ignorance than knowledge" (Daniel, 2021).

Companies deploy alternative forms of evaluation to face these uncertainties and thus become advocates of compost as an agronomic quality product. These evaluations are often complemented by other labels, assessments, and numerous scientific partnerships. For example, the production of compost from baby diapers is subject to a scientific consortium to determine if the produced compost meets the required quality for use as a fertilizer and agricultural amendment. The quality of compost can also be assessed through experiments in partnership with urban agriculture organizations.

These alternative evaluations demonstrate that a "good" compost is not only one that complies with regulations, but also a product that guarantees agricultural and ecological quality. For instance, Cathy, the founder of a branch of Bikompost, explained in an interview all the steps taken to ensure the quality of the compost, well beyond regulatory requirements, which only demand that compost be approved by the AFNOR NFU 44-051 standard to be sold. When the compost produced by the Bikompost association did not initially obtain the standardization, Cathy engaged in a scientific partnership with a support organization to address the issue:

> We decided to invest in composting with a scientific partnership, with an organization called APESA, which accompanied us for a year in optimizing our composting practices. Because we handle quite large quantities, and we do everything by hand, but we still wanted to do things properly. (*Ibid.*)

[8] My translation.

This support led them to work with two agronomists for a year and carry out a series of measurements—oxygen, temperature, pH, humidity—which then allowed them to feel like they had "good control" over the composting process. After this initial optimization of the composting process, Cathy implemented additional analyses to ensure that the compost could be used in organic agriculture, which required official certification, separate from the norm. More generally, she emphasizes her desire to work with "calls for projects" and "consortia," both with agronomists and with sociologists on waste issues—a taste for research that she justifies as a way to "stay in touch" with her former career as a researcher in a large public company, after earning a doctorate in mathematics.

The scientific partnerships are sometimes extended through experiments in collaboration with market gardeners who use the compost and become advocates for promoting high-quality compost as an agricultural fertilizer. This is the case, for example, with a diaper composting project implemented by Magicompost, which is accompanied by ADEME (French Environment and Energy Management Agency) and researchers from an engineering school. While at the time of the interview, the main goal is to see if the compost from compostable diapers aligns with the NFU 44-051 standard, Rebecca, who leads the project, also envisions "additional analysis" with research partners to conduct studies and test future uses. They may even try to influence the rewriting of standards in light of the ban on the sale of composts from Mechanical–Biological Treatment (TMB) processes and the emergence of various compostable products. The defense of compost quality also involves political lobbying efforts to "reassure" and "show everyone" that these new composts have "benefits for soils" and to influence the evolution of standards and public policies.

The understanding of the cause evolves through contact with the materials and technical processes. While the founders of these companies are primarily waste management professionals, they become advocates for compost as a high-quality organic soil amendment. Despite these similarities, what constitutes a good compost is a subject of debate among different companies, which interpret the cause in various ways.

12.4 Beyond the Cause, a Variety of Implementations

The companies studied are indeed working toward a common cause: turn food waste into rich soil. However, they interpret and put this cause into practice in very distinct ways. What constitutes good compost(ing) takes on different meanings from one company to another and can be related to certain political and economic divisions. First, I will show that Bikompost advocates for ecologically exemplary practices at the expense of regulatory and economic concerns. Secondly, I will demonstrate that Magicompost prioritizes commercial interests, even regarding compost, which affects the assessment of compost quality.

Prioritizing Compost's Agronomic and Ecological Quality Over Economic Resources

For Bikompost, offering an ecological service means paying attention to the ecological consequences of the means employed. Instead of prioritizing efficiency, the company favors collection and composting methods considered virtuous. In this regard, collection is done by bicycle, and waste processing is not mechanized. Prioritizing ecological exemplariness has several consequences. These technical choices involve managing only small quantities of waste. Furthermore, ecological exemplariness leads to makeshift solutions that are incompatible with economic productivity logic.

Collecting waste by bicycle, for example, involves transforming a tool originally designed for personal transportation into a technical device capable of carrying heavy loads. Collecting by bicycle thus entails diverting this object from its primary function since, initially, "no bicycle in the world" is designed to "pull 300 kilograms," as emphasized by Bastien, the equipment manager at Bikompost. Bicycle collection, therefore, requires equipping the bicycles with suitable collection accessories, such as a bicycle trailer. This adaptation work is challenging due to the limited use of bicycles as professional tools. When Elsa founded the company in 2015, "there was nothing in France for waste collection by bicycle." With the support of a collective of companies from various sectors that all use bicycles as a working tool, Elsa discovered the existence of trailers adapted for professional bicycle use. It was also necessary to find collection bins that were suitable for the trailer: these were not based on the model of conventional bins but were rectangular bins of 50 liters, about 40 centimeters high, which had to be carried by hand. With limited institutional oversight, the implementation of bicycle collections is marked by trial and error and uncertainties about the proper functioning of the tools.

Like bicycles, manual composting is characterized by a series of improvisations, including the repurposing of materials and objects originally not intended for such use. For example, weighing the sorting buckets is done using a hook scale, such as those used in fishmongers. Materials are also collected to test alternatives to commercially purchased wood chips, both to prioritize ecological alternatives and reduce costs. The company plans to recover the substrate used by a market gardener to grow mushrooms—a "mixture of straw and cereals"—or the wood scraps from a cabinetmaker. Finally, technical objects are created specifically for the activity, such as a manual sieve for compost. The chosen tool is used manually—a handle allows the cylinder with a mesh grid to rotate, sieving the compost. Its operation is not optimal: Axelle, a Bikompost employee who conducted the sieving the day before we met, told me that using the sieve had spread compost everywhere—"we almost took longer to clean up than to sieve," she concluded with a laugh.

The choice of ecologically exemplary technical processes is particularly marked by trial and error. In this regard, Bikompost's technical directions are incompatible with a goal of maximum productivity. The time required for technical improvisations and the relative effectiveness of the devices implemented necessarily break away from a logic of economic and technical performance.

This positioning has consequences for the perception of compost, and more broadly, for soils. Indeed, from the perspective of ecological exemplariness, compost is primarily seen as an agronomic product to be carefully produced, with economic value being set aside. In some local branches of Bikompost, the choice has been made to compost directly at a farmer's place: in exchange for a dedicated composting space, the compost produced is given to the farmer for free. One Bikompost branch even completely abandons the collection activity, even though it is the most financially rewarding. The founders of this branch choose to focus on compost production and training residents in local composting, thus prioritizing compost quality over economic productivity. This choice marks a shift from a waste management professional stance to a composting professional stance.

Toward a Commercial Vision of Soils and Compost

Magicompost, on the other hand, aims to offer an ecological and alternative service on a larger scale, capable of handling the waste generated by urban households. In practice, this leads to a collection service using trucks. The collected waste is then processed at mechanized sites on the outskirts of urban areas. These large sites, capable of processing between 700 and 2000 tons per year, are equipped with several machines that facilitate and accelerate the composting process. Firstly, the waste is dumped onto a sorting table using a mechanical bin lifter. Once sorted manually, the waste is transferred to a grinder and then a mixer, which mixes the food waste with the woodchip. It is then composted for several weeks in an electromechanical composter, designed to speed up the composting process. The technical processes are therefore marked by motorization and mechanization.

However, this does not mean that the directors consider they have abandoned an ecological perspective. On the contrary, they believe that the quantity of waste collected is the only way to have a real "ecological impact," to use their words. The argument of quantity mitigates the ecological consequences of the means used, such as the pollution generated by trucks. In their views, the large quantities would justify the shift to an industrial mode, presented as an obvious choice.

These technical choices have consequences for the compost produced. In a perspective of economic productivity, it is primarily seen by the company as a commercial product. These commercial orientations significantly influence the evaluation of compost quality, as evidenced by the debates around damaged compost bags. During a day of work at a Magicompost treatment site, we came across more than 15 boxes filled with two-liter compost bags. We opened them to check the condition of the bags, which turned out to be in poor condition. A number of bags were soiled on the outside—marked with brownish stains—and on the inside—mycelium, a type of white fungus, had proliferated. A small group then gathered around the bags to debate the quality of the compost—was it moldy?—and to decide if it could be sold. The group consisted of three members of the commercial department and the operations manager of another composting site, the only one with a degree in agricultural engineering, in addition to my role as a participant observer. To assess the quality of the compost, visual evaluation played a significant role, rather than relying on agronomic or health-related arguments. To determine if the compost could be delivered to distributors and sold in small bags, the assessment was primarily visual: the exterior of the bag was examined to see if it was soiled or not. If the bag showed no signs of soiling on the outside, it was placed back in the box for delivery to distributors. Bags with exterior stains were opened to visually assess the amount of mycelium on the top of the compost. When the compost was dotted with small white spots—the mycelium—an olfactory evaluation was performed, with some asking others: "what does it smell like to you?" Opinions varied significantly: some thought it smelled "moldy," while others thought it smelled like "the forest floor." A tactile evaluation also complemented the visual and olfactory assessments. Throughout the process, those present continually exclaimed about the heat generated by the compost bags: "it's hot!," "it's alive in there!". The question of the agronomic quality of the compost ultimately remained unresolved in favor of a commercial definition of quality. Significant commercial losses were at stake for the company: "there's 3.000 euros worth of merchandise in there" sighed Anna, in charge of compost sales at the time. Thus, most clean bags on the outside were placed back in boxes without even checking the condition of the compost inside. In addition, the two saleswomen were considering the strategy to adopt in case of a problem with a distributor and/or a customer. They were developing an argument to use with distributors in case of complaints: "if there are complaints [from distributors], we need to explain that it's a living product, that it's normal, even beneficial for plants." They also discussed the possibility of allowing customers who wished to exchange their product. However, the presence of mycelium as a guarantee of agronomic quality is far from certain. This idea is debated within the company itself. Some of those present claimed that mycelium is not harmful and attributed any customer hesitation to a problem of "education" and ignorance about gardening. Others, in contrast, pointed out that while mold may form "naturally," it is not necessarily beneficial—"if you leave wet shoes in a corner, mold will also form." Damien, the technical manager, suggested that mycelium might be a sign of insufficient stabilization of the compost at the time of packaging, more indicative of an error in the production process than a natural and beneficial consequence.

In addition to this commercial view of compost, Magicompost also adopts a marketing approach to the cause of soil. In 2020, the company decided to rethink its "brand identity" and emphasize its role as a defender of soils. This "brand strategy" involves organizing events such as the screening of a documentary film dedicated to soil issues and a conference on the same topic with various guest speakers. More than establishing a direct connection with agricultural professionals, Magicompost positions itself as an advocate for the cause of soils. The company engages in educational efforts toward households, who can, for example, visit the composting site. This educational work is closely tied to commercial goals since households are also the main customers for the compost, mostly sold in small bags.

Unequal Legitimacy: Conflicts Around the Cause

The two implementations of the cause not only differ in waste management and compost production but are also unevenly endorsed by local governments.

Firstly, Bikompost's technical choices, such as bicycle collection and manual composting, are disapproved of. Although the company presents itself as a cycling professional—investing in high-end bicycles, specialized trailers, and hiring a dedicated fleet maintenance employee—its collection and processing capacities are deemed insufficient by public officials. Their technical choices are also criticized and described in some cases as "regulatory failures." The relationship between Bikompost and public decision-makers consists of many tensions, which are also related to the fact that Bikompost presents its activity as a "public service." In Bikompost founders' views, their business should be supported by public institutions regarding its public interest. Public decision-makers strongly contest this idea, by arguing Bikompost is a private business—no matter how socially useful it can be.

In contrast, Magicompost obtains a dominant position in the emerging economic field thanks to its technical and economic choices. For example, the business secured several public waste collection contracts. This dominant position is then the result of a persuasive effort directed at public decision-makers, involving sending letters to elected officials and organizing regular meetings or site visits. The goal is to convince local authorities of Magicompost's ability to handle large quantities of waste, as emphasized in the company's motto—"we are not small!" Despite its humorous tone, this motto underscores the pursuit of greatness, both literally—the capacity to absorb significant amounts of waste—and figuratively—being perceived as a serious and credible company by public authorities.

While these more strategic stances toward public decision-makers can explain the unequal positions of the two companies, the individual trajectories of the founders also vary. On one hand, Magicompost's persuasive work is facilitated by the economic and social resources available to the company's leaders, mostly coming from the economic strata of the middle and upper classes. In addition to possessing significant educational capital due to their training at prestigious business and engineering schools, they have a substantial social network. Maximilien, one founder of Magicompost, embodies this social capital, both through the charisma attributed to him within the company and through the interview with him serving as a showcase of his social capital. He continually references various friends or acquaintances in leadership positions. Moreover, Magicompost's leaders largely draw on their prior professional socialization in strategic roles and entrepreneurship. Among the ten founders of the various branches, four previously held leadership positions in large companies, and five had started their own businesses or participated in the development of a start-up. The professional skills they acquired are then put to use in the company, which translates into significant financial resources compared to other companies in the sector. Magicompost has raised funds through investors, enabling it to make technical investments and hire more employees.

In contrast, Bikompost's founders have fewer educational resources. They completed shorter or less socially valued studies, such as language programs. Having less frequently held leadership positions before starting their business, they rely less on prior professional socialization. However, these founders possess certain specific resources, often related to their activist commitments. The local activist network, for example, provides volunteer labor and logistical and publicity support. Bikompost's activist positioning is therefore ambiguous. On the one hand, its activist stance is seen as a sign of unprofessionalism, disqualifying the company. On the other hand, activism provides certain resources, even political support in some cases. Bikompost receives support from a municipal official responsible for the circular economy, who is a member of the local zero-waste group, contradicting the recommendations of the local governement's technicians.

Examining the practical implementations of the cause reveals conflicts. While each organization appropriates the cause of compost differently, these different models are unequally legitimized by public decision-makers, who value the professional waste management model over that of soil advocates.

12.5 Conclusion

This chapter allows for further analysis regarding the non-emergence of soils as a "public issue." First and foremost, public action is primarily focused on managing the significant quantities of waste generated. Despite the promotion of a "circular economy" and a "return to the land" for food waste, public policies primarily revolve around waste management. Nevertheless, entrepreneurs in the waste management sector gradually embrace the cause of soils and become more acquainted with it. Learning how to transform organic matter into high-quality compost encourages broader knowledge about soil life.

Nonetheless, the practical implementation of this cause within companies significantly differs due to their technical choices. Two practical visions clash: one emphasizing ecological exemplarity and the other emphasizing ecological productivity. These differences have consequences for compost production priorities. On one side, the emphasis is on the agronomic quality of compost, while on the other side, it is geared toward commercial purposes. If the cause of soils struggles to emerge, its emergence is also problematic and fraught with conflicts that reflect heterogeneous ecological perspectives.

My research thus highlights the ambivalence and conflicts surrounding the "return to the land" cause for food waste, as seen through the analysis of the emerging market for food-waste collection and composting. Organic soil fertilization, however, also raises issues for market gardeners. This investigation could be extended to examine the fate of these organic materials and the reception and utilization of these materials in the agricultural world.

Competing Interests and Acknowledgments This work was carried out with the financial support of EHESS (Ecole des Hautes Etudes en Sciences Sociales).

References

Barles, S. (2009). Urban metabolism of Paris and its region. *Journal of Industrial Ecology, 13*(6), 898–913. https://doi.org/10.1111/j.1530-9290.2009.00169.x

Barles, S. (2016). *L'invention des déchets urbains : France 1790–1970*, Editions Champ Vallon.

Becker, H. S., & Richards, P. (2007). *Writing for social scientists. How to start and finish your thesis, book or article*. Chicago University Press.

Bereni, L. (2021). The women's cause in a field: Rethinking the architecture of collective protest in the era of movement institutionalization. *Social Movement Studies, 20*(2), 208–223. https://doi.org/10.1080/14742837.2019.1679107

Daniel, F.-J. (2021). Normalisation et « retour au sol » des résidus organiques. *Revue d'anthropologie des connaissances, 15*(4). https://doi.org/10.4000/rac.23884

Evans, D. (2012). Beyond the throwaway society: Ordinary domestic practice and a sociological approach to household food waste. *Sociology, 46*(1), 41–56. https://doi.org/10.1177/0038038511416150

Fournil, J., Kon Kam King, J., Granjou, C., & Cécillon, L. (2018). Le sol : Enquête sur les mécanismes de (non) émergence d'un problème public environnemental. *VertigO - la revue électronique en sciences de l'environnement, 18*(2). https://doi.org/10.4000/vertigo.20433

Granjou, C., Higgin, M., & Mounet, C. (2020). Le compostage, entre réduction des déchets et domestication du pourrissement. *Revue d'anthropologie des connaissances, 14*(4). https://doi.org/10.4000/rac.11873

Gusfield, J. R. (1997). *The culture of public problems: Drinking-driving and the symbolic order*. Routledge.

Henry, E. (2007). *Amiante, un scandale improbable : Sociologie d'un problème public*. Presses universitaires de Rennes.

Krzywoszynska, A. (2012). 'Waste? You mean by-products!' from bio-waste management to agro-ecology in Italian winemaking and beyond. *The Sociological Review, 60*(2), 47–65. https://doi.org/10.1111/1467-954X.12037

Lehec, E. (2018). *La remise en cause des services urbains en réseau : le cas du compostage des déchets en pied d'immeuble à Paris*. Université Paris 1.

Lupton, S. (2017). Markets for waste and waste–derived fertilizers. An empirical survey. *Journal of Rural Studies, 55*, 83–99. https://doi.org/10.1016/j.jrurstud.2017.07.017

Maslen, S. (2015). Researching the senses as knowledge. *The Senses and Society, 10*(1), 52–70. https://doi.org/10.2752/174589315X14161614601565

Maslen, S. (2016). Sensory work of diagnosis: A crisis of legitimacy. *The Senses and Society, 11*(2), 158–176. https://doi.org/10.1080/17458927.2016.1190065

McCarthy, J. D., & Zald, M. N. (1977). Resource mobilization and social movements: A partial theory. *American Journal of Sociology, 82*(6), 1212–1241.

Meulemans, G., & Granjou, C. (2020). Les sols, nouvelle frontière pour les savoirs et les politiques de l'environnement. *Revue d'anthropologie des connaissances, 14*(4). https://doi.org/10.4000/rac.14027

Monsaingeon, B. (2017). *Homo detritus : Critique de la société du déchet*. Éditions du Seuil.

Neveu, É. (2017). L'analyse des problèmes publics. Un champ d'étude interdisciplinaire au cœur des enjeux sociaux présents. *Idées économiques et sociales, 190*(4), 6–19. https://doi.org/10.3917/idee.190.0006

O'Brien, M. (2012). A 'lasting transformation' of capitalist surplus: From food stocks to feedstocks. *The Sociological Review, 60*(2), 192–211. https://doi.org/10.1111/1467-954X.12045

Puig de la Bellacasa, M. (2014). Encountering bioinfrastructure: Ecological struggles and the sciences of soil. *Social Epistemology, 28*(1), 26–40. https://doi.org/10.1080/02691728.2013.862879

Vatin, F., Callon, M., & Desrosières, A. (2013). *Évaluer et valoriser : Une sociologie économique de la mesure*, Presses Universitaires du Mirail-Toulouse.

CHAPTER 13

Mechanical Tillage in Organic Viticulture and Its Effect on Soil Life

Jérôme Gidoin

The rise of organic[1] viticulture is tied to the exclusive use of mechanical tillage. This shift can be seen as a "return to the soil," a response to the widespread use of synthetic chemical pesticides. It is a trend that underscores soil health and the relationship between tillage and soil life. This discussion seeks to examine that relationship more closely. Before we dive deep into the core of the matter, it is important to briefly set the historical and axiological context of the resurgence of mechanical tillage in organic viticulture before detailing its essence from both a technical and practical standpoint. Next, I will explore the current discourse surrounding tillage and its connection to the concept of "living." It seems we are currently experiencing a revival in the significance of understanding and

[1] For the sake of convenience, I will include under "organic" viticulture not just officially certified organic viticulture but also biodynamic viticulture and viticulture following the "natural method."

J. Gidoin (✉)
Université Paris-Descartes, Paris, France
e-mail: j.gidoin@orange.fr

AnthropoVino, Paris, France

© The Author(s), under exclusive license to Springer Nature Switzerland AG 2025

O. Labussière et al. (eds.), *Back to the Ground*,
https://doi.org/10.1007/978-3-031-88888-5_13

managing soils in viticulture, signaling potential shifts in future methodologies and perceptions. I intend to contrast two fundamentally opposed perspectives: One staunchly advocates for tillage as an essential foundation for sustainable soil management in organic viticulture, while the other proposes minimizing soil disturbance to protect the subterranean microbiological ecosystem, especially the earthworms that are instrumental in naturally enhancing soil fertility. From these conflicting viewpoints, I aim to outline two distinct relationships with the soil and two approaches to action or intervention: one that controls and "steers" the soil and another that seeks to foster a sense of otherness with the soil as a habitat for biodiversity. This analysis serves three purposes. First, it critically examines the connection between mechanical tillage and soil life. Second, it identifies the "naturalistic" influence of mechanical tillage on the soil. Third, it argues that the alternative to mechanical tillage is not inaction or simply "letting nature take its course," but rather, engaging with the soil in different ways. To support this notion, I will refer to a specific agricultural practice that reduces soil disturbance: the use of permanent cover crops by the winemaker Sébastien David in Saint-Nicolas de Bourgueil. The ultimate goal is to move beyond the dichotomy often accepted in organic viticulture, whether consciously or subconsciously: mechanical tillage (human intervention; culture) *vs.* non-tillage (lack of intervention; nature).

To bolster my argument, I will partly rely on ethnographic data I gathered in 2017–2018 during my year of professional training in organic viticulture. In my pursuit to dedicate a segment of my life to winemaking alongside anthropology,[2] I was fortunate enough to receive funding from the Conseil Régional du Centre-Val de Loire for a specialization certificate in organic viticulture (CS VITIBIO), entitled "Organic wine production, marketing, and processing" (*Conduite de la production viticole en agriculture biologique, commercialisation et transformation*)," at the Lycée Agricole et Viticole d'Amboise. This program with a uniquely organic focus is the first of its kind in France and was established in 2014

[2] I have been a microscale winemaker for five years, cultivating a one-hectare vineyard in Cinq-Mars-La-Pile, Indre-et-Loire.

to nurture the region's burgeoning organic viticulture scene.[3] Leveraging this rich blend of theoretical (600 hours of classroom learning at the training center) and practical experiences—especially during fortnightly internships (420 hours) at a biodynamic vineyard in Chinon—I conducted an ethnographic study (observations, interviews, and discussions gathered during lessons or talks; I always had a field notebook on hand and used my smartphone to take photographs). Here, for the technical part, I will reference my course content[4] on mechanical tillage, which essentially serves as an ethnographic data corpus on the teaching of tillage in organic viticulture training. My aim is not comprehensive coverage but to focus on what I believe is crucial to remember in retrospect. When discussing the contrasting types of relationships with the soil, I will share insights from Dominique Massenot, a renowned and influential figure in biodynamic viticulture. Massenot, an ardent supporter of soil tillage, offered us his sharp critique of the agroforestry method (and agroecology[5] more broadly) in relation to viticulture during his lecture on understanding soil in biodynamics. Lastly, the concluding part draws from a lengthy conversation with the winemaker Sébastien David in 2022, which occurred outside the scope of my training.

13.1 Mechanical Tillage: 'A Return to the Soil'

Mechanization involves the use of machinery or mechanical tools, which are not necessarily powered, to alter the dynamics of human (i.e., manual) or animal labor. This could mean, for example, converting a straight-line motion into rotation, amplifying speed and power, or even creating a

[3] In 2011, the Agence BIO Centre reported that of the region's 22,000 hectares of vineyards in Indre-et-Loire and Loir-et-Cher, 9.5% were dedicated to organic viticulture, and a 19% increase in area between 2010 and 2011. By 2021, the Centre region had markedly distinguished itself, with 28.4% of its vineyards being organic, compared with 20.5% nationally.

[4] This 30-hour course, both theoretical and practical (involving operating a vineyard tractor fitted with various tillage implements), was led by the vineyard manager of a renowned 20-hectare estate in the Vouvray appellation.

[5] Agroecology can be defined as "the application of ecological principles to the design and management of sustainable agricultural systems." This notion "encompasses a broad range of critiques in action against the modernist paradigm," specifically regarding "the socio-ecological risks associated with the prevailing agronomic model" (Charbonnier, 2022: 302–306).

leverage effect. On the viticultural landscape of France in 1945, there were hardly 1000 tractors; traditional soil management was primarily conducted with a vineyard plow and a vine row cultivator, relying mostly on animal draft. Following World War II and bolstered by support from the Marshall Plan, there was a surge in motorization, leading to what could be termed widespread "tractorization" from the 1960s onward. By 1958, the number of tractors had increased approximately a 100-fold since 1945. Alongside the motorization of viticulture in the 1950s, vine-growing adopted more industrial practices, and there was a significant uptick in the use of mineral fertilizers and chemical industry inputs (created by combining chemical elements). The primary arguments in support of this technology include time-saving (herbicides are increasingly used as substitutes for mechanical weeding, which is a significant aim of tillage), reduced hardship, enhanced productivity (partly owing to the use of mineral fertilizers), improved distribution of labor (especially because the efficacy of systemic treatments against diseases is less variable than that of prophylactic treatments), and a reduction in costs associated with equipment and labor (decreased tillage leads to less tractor use).

The organic farming movement emerged in the 1970s[6] to counter the widespread use of pesticides. In the realm of viticulture, this movement was more delayed because vines and wine were not necessarily seen as immediate concerns; the primary goal was to provide healthy food for humans while taking care of resources. In viticulture, however, treatments are particularly intensive and systematic, mainly due to high yields, which encourage the development of disease. "Organic" viticulture began to take shape in the 1980s and 1990s and really took off in the 2000s.[7] It offers an alternative perspective on production; its approach is holistic, taking into account production, environmental, and health aspects.

In this context of opposition to viticulture reliant on the chemical industry, I will reframe the resurgence of mechanical tillage in "organic" viticulture. Here is what one can read in the introduction of a technical sheet that served as a basis for one of my CS VITIBIO courses specifically dedicated to tillage ("Fiche Tech' Sol, travail du sol," ATV, Association Technique Viticole 49, February 2016):

[6] Established in 1972, the International Federation of Organic Agriculture Movements (IFOAM) has promoted global awareness of agrobiology.

[7] In 2010, there were 50,000 hectares; this increased to 60,000 hectares in 2011, marking a 20% rise within a year and accounting for 7% of all French vineyards.

> To address social and environmental challenges, including preserving human health and the quality of our waterways, the reduction in herbicide use will likely accelerate in the coming years. (...) When performed under appropriate conditions, mechanical tillage offers an alternative that marries efficiency with agronomic value.

The primary concern here is the impact of chemical weeding on soil, which leads to its degradation and contamination (with adverse geochemical reactions creating molecules often more persistent and hazardous than the originals) and affects all living beings.

This "return to the soil" also grants autonomy to the winemaker, who is no longer dependent on the services of a pesticide provider. This shift is exemplified by Emmanuel Caslot, a chemist by training and the elder brother of one of my classmates from LPA Amboise. In an interview published in the quarterly "Le Vin ligérien" (2012: 42), he shared his experiences working at the renowned family estate La Chevalerie, located in Bourgueil. He and his two sisters subsequently assumed control of the estate in 2014 following their father's death:

> I started a training course for adults at the Montreuil Bellay agricultural school. There, I met individuals like Mark Angeli. Back then, there was scant discussion of biodynamics. Two years later, we stopped using herbicides on the estate. We fired the guy who came to sell us his chemicals, and we were on our own.

In the Practical Guide to Organic Wine ("Guide pratique du vin bio"), a seminal text for those undergoing professional training and one that I regularly consulted during my studies, it is stated that tillage "is the quintessential aspect of organic viticulture, which forbids the use of herbicides" (Natoli, 2013: 47). Indeed, tillage differentiates organic from conventional viticulture, although it is important to note that tillage is also performed in conventional practices, albeit not exclusively.[8] To put it another way, at the risk of being repetitive, the major difference lies in the use or non-use of synthetic pesticides in two fundamental areas: on the one hand, weed and grass control, and, on the other hand, soil fertilization—pitting mechanical weeding against herbicides organic and mineral

[8] By 2016, half of the French vineyards employed a combination of chemical and mechanical weeding methods.

fertilizers of natural origin against predominantly mineral synthetic chemical fertilizers. The dichotomy of "organic" *vs.* "chemical" or "en bio" *vs.* "en chimique," as commonly articulated by various stakeholders in the viticulture sector (e.g., in the trade press aimed at professionals, but also in the vernacular of "organic" viticulture practitioners), thus marks a distinction on both technical and value-based levels.

I would also note that this dichotomy has led to a common association, which I have often observed. For example, my internship supervisor Pascal Lambert, a winemaker, associates "organic" with "living soil." Conversely, "conventional chemistry" is linked to "dead soil." This distinction has been popularized by agronomists Lydia and Claude Bourguignon. However, although "organic" and conventional vineyard plots can be easily differentiated through tillage practices, particularly at the level of the vine mound ("cavaillon," the earthen mound at the vine's base), the stark binary contrast of "living soil" *vs.* "dead soil" is contested by most soil microbiologists.[9] Preserving soil life has even been used as a justification by proponents of herbicide use in conservation agriculture (Goulet, 2008). The real problems caused by the use of synthetic chemicals are the contamination of soil and groundwater, the destruction of biodiversity,[10] and the risks posed to the health of all living organisms.

13.2 Some Fundamental Technical Aspects of Mechanical Tillage in Organic Viticulture

Compared with chemical weed control, mechanical weeding entails an additional cost of between 30 and 50%; it also means a higher wage bill and a tenfold increase in overheads, including the purchase and maintenance of specific equipment, fuel, etc.

[9] Most herbicides, particularly those targeted by organic viticulturists, are less harmful to non-target living organisms than other pesticides like insecticides, copper-based fungicides, nematicides, and molluscicides, which can decrease soil biodiversity by more than half. However, while herbicides' lethal impact may be less significant, they impair the development of individuals across generations, for instance, by diminishing fertility (Feller et al.: 141).

[10] The notable enhancement of soil biodiversity associated with organic agriculture (elevated microbial biomass, augmented bacterial and fungal activities, increased diversity of nematodes, earthworms, insects, and arthropods) is primarily attributed to the reduction or absence of pesticide treatments (Feller et al.: 146).

There are different approaches and purposes to mechanical tillage, which will determine the depth and intervals (systematic or otherwise) of intervention. It is also important to differentiate tillage in organic viticulture from plowing. While plowing remains axiological for a large proportion of farmers, it is less and less recommended and practiced in organic viticulture, because penetrating deeply and turning over the soil disrupts the soil structure and its natural layers.

I will concentrate on what I consider to be the three main goals[11] of tillage (after planting), without going into the technical specifics of the various tools employed, nor their particular applications and rationales. The first objective of tillage is to control weeds, which compete with the vines for water and nutrients (particularly nitrogen). The second is to loosen the soil in order to enhance root penetration and the structuring of the tilled soil. The third goal is to sustain or augment soil fertility. For each of these goals, there are two main areas of work: within the vine row, around and between the vines, and across the inter-row spaces.

Weed control within the vine rows can be effectively conducted through the process of under-vine plowing. This involves the use of an under-vine plow that slices deep into the topsoil at and between the vine stocks, then flipping the excavated strip of soil (toward the vine stocks) and cutting through the roots of the weeds. This procedure can be considered a form of highly localized plowing. The reverse operation to under-vine plowing is referred to as "ridging." An alternative, less invasive technique employs horizontal inter-row blades ranging from 40 to 60 centimeters in length to horizontally slice a strip of soil a few centimeters deep without displacing the soil, thereby desiccating the above-ground parts of unwanted vegetation. This method of weeding is comparatively superficial and functions similarly to hoeing. The term "flat tillage" contrasts with plowing in that it does not alter the soil's topography.

Regular hoeing with inter-row blades is typically enough to prevent the establishment of weeds such as brambles, ivy, and couch grass, among others. However, if these weeds are firmly established, hoeing may paradoxically stimulate their regrowth. In such instances, under-vine plowing

[11] The purpose of tillage in viticulture differs somewhat depending on whether it is carried out before or after planting. Before planting, the primary focus is on optimizing conditions to facilitate root development in young vines: loosening compacted soil and extracting roots that could potentially transmit diseases, such as root rot.

becomes advisable since deeply slicing the roots and flipping the soil facilitates more efficient weed removal. If there is a significant weed infestation, it becomes technically challenging and economically impractical (owing to the high cost for minimal outcome) to operate at depths shallower than 10 centimeters. Additionally, the flora adapts to cultivation practices; for instance, if a vine mound is consistently managed with inter-row blades, it may gradually succumb to rhizomatous weed invasion. Thus, the ideal solution is to diversify practices through the utilization of various cultivation tools.

In the inter-row spaces, weed control is generally achieved by using hoeing techniques. Tools equipped with tines offer a viable alternative to plow-based tillage for mechanical weeding. These tines scratch the soil approximately 12 centimeters deep, uprooting weeds that are then superficially reincorporated into the soil.

Mechanical weeding also serves to loosen the soil—that is, to decompress and aerate it. Hoeing breaks up the compacted surface crust created by rainfall and the repeated passage of machinery, thereby ensuring an optimal water/air gas exchange and managing evaporation. The resulting coarse fragmentation of the surface layer enhances water absorption and percolation at depth.

The goal of soil loosening is similar to that of soil fertilization. The cultivation of inter-row spaces maintains soil fertility by reintroducing organic matter, including sugars and nitrogen, to the surface layer to nourish the microorganisms responsible for structuring the soil into aggregates. Furthermore, soil aeration facilitates the biodegradation and mineralization of organic matter, thus altering the kinetics of nitrogen release from the soil. Hoeing may also serve as a preparatory step for soil fertilization, whether through the addition of organic or natural mineral fertilizers or for the sowing of cover crops (green manures).

From a technical and practical standpoint, the relationship between mechanical tillage and soil life fundamentally hinges on two aspects. First, tillage helps control weed growth and fertilize the soil without the risk of groundwater contamination, thereby offering organic viticulture a more environmentally respectful alternative to conventional methods. Second, by aerating the soil, tillage stimulates subterranean microbial life, accelerating the biodegradation and mineralization of organic matter.

13.3 Debate on the Effects of Tillage on Soil Life

In the realm of "organic" viticulture, there is currently no pressing desire to challenge the fundamental principles of tillage, though it is worth noting a recent shift toward a less systematic approach, partly due to global warming. Nonetheless, the purported benefits of these practices are now under scrutiny.[12] For instance, A. Canet, an expert in the application of agroforestry to viticulture, asserts that systematic tillage depletes soil quality in both agriculture and viticulture. This depletion occurs initially through erosion, which leads to significant soil loss because of gullying. Second, such practices hinder the storage of carbon that is otherwise facilitated by photosynthesis in very significant proportions.[13]

Importantly, carbon storage in soil not only augments the organic matter content but also enhances the soil's capacity to retain water. This, in turn, stimulates microbial activity within the soil and helps reduce the reliance on fertilizers and the emissions of nitrous oxide, which is a potent greenhouse gas. It is estimated that in France, agricultural tillage discharges half of the carbon that should be stored in the soil every year across all types of cultivation. Therefore, it is advisable to draw inspiration from the resilience and self-regulating biodiversity of forest soils and to embrace crop diversification as a key principle. This might include strategies such as planting trees and establishing grazing areas during winter months. These concepts are gaining traction and beginning to influence some "organic" winemakers, details of which will be discussed later. Indeed, these ideas are contributing significantly to the ongoing debate over the sustainability of viticultural soils.

The current level of interest (or apparent disinterest) in earthworms[14] provides a particularly telling lens through which to view the epistemic tensions related to the impacts of tillage on soil life:

[12] For a genealogical perspective on this shift toward a new understanding of soil in terms of threatened biodiversity, see Fournil et al. (2018).

[13] This is an ecological argument that views the soil in its entirety and not merely its productive aspects.

[14] An increasingly popular method for assessing a soil's biological health involves collecting earthworms from a small patch of soil (about 25 cm^3) by using an irritant like vinegar or mustard.

> Every time a tool is used, the soil suffers! This process disrupts the "mother house" of earthworms, which are pivotal to soil health. (…) The soil acts as a lung, breathing in air and fixing nitrogen. (…) Whenever the top 3 centimeters of soil are disturbed, earthworm burrows, which are crucial for maintaining soil fertility cycles, are destroyed. (A. Canet,[15] agronomist and agroforester, former president of the French Association of Agroforestry)

For agro-pedologist D. Massenot, however, this perspective is folly. He advocates vigorously for tillage as an essential and indispensable component of "organic" viticulture and says that vineyard soil bears no resemblance to forest soil:

> Mulch is fine for strawberries and cucurbits but can't be universally applied, especially not in viticulture. Otherwise, you'd need tons of manure per hectare, which would ultimately degrade the soil[16]; this approach is fundamentally anti-agronomic! (…) There's a lot of nonsense about organic matter, particularly the belief that having more necessarily increases fertility. Humus consists of very stable organic matter and is, thus, minimally available for crop use. It is preferable to use small quantities of soluble organic materials, which are rich in sugars and readily available nitrogen. (…) There's also a lot of nonsense about earthworms, whose presence has no agronomic value; without them, soil can remain clumpy due to other microorganisms, maintaining adequate conditions for microbial life. (…) Without tillage, yields are inevitably lower or even nonexistent, and this, in turn, causes problems in the cellar for wine fermentation, notably due to the lack of nitrogen in the must, which generally leads to organoleptic deviations. (…) Agronomy is back in viticulture today. However, there has been a move away from chemical weed control to permanent grass cover, which is not advisable. Adaptation and modulation according to soil type are necessary. Vineyard soils have unique characteristics that must be acknowledged, regardless of the preferences of permaculture and agroforestry proponents.

[15] Data was collected by journalist Juliette Cassagnes in her article "Un jour, il faudra arrêter le travail du sol dans la vigne," dated 31 January 2014, available at www.vitisphere.com.

[16] This is likely due to the acidification of the soil linked to the mineralization of organic matter.

These two antagonistic points of view provide us with some insightful arguments concerning the debate over tillage *vs.* less (or even no) tillage. First of all, it is important to note that Massenot's approach is fundamentally based on the "Hérody method," of which he is a prominent advocate. This method is prevalent in organic agriculture and has been trialed in several developing countries. It views soil life primarily through the lens of microbial activity that decomposes organic matter and organizes soil into aggregates. When soil functions effectively, its aggregation[17] is optimal, thus creating a structure that supports robust root development. Hérody proposes a practical methodology for managing cultivated soils by integrating physicochemical analyses with agropedological field observations. He argues that tillage is a vital cultural practice as it effectively conditions the soil's microbial environment. By modifying the circulation of air and water within the soil, tillage enhances the soil's microbial activity, thereby indirectly influencing soil structure, with tillage tools merely causing fragmentation.

According to Massenot and Hérody[18]: "Earthworm activity is better suited to the processing of mulch, that is, uncultivated soils. In cultivated soils, it should only supplement the 'root/microbe' interaction, never replace it" (Hérody: 24). They contend that deep root development efficiently enriches the soil with transformed organic matter,[19] a process that relies on human intervention through tillage.

The term "mulch" is used here with a pejorative connotation to present a reductive view of forest soil. This perspective also underlies the argument concerning the "tons of manure per hectare." From an agroecological standpoint, the true value of forest soils lies not merely in the abundance of organic matter but rather in the density and depth of their root networks, as well as the intricate biodiversity that emerges

[17] Aggregation involves the creation of chemical bonds and the binding of chemical elements. Various "binders" are employed, including mineral, organic, and organo-mineral types, with the clay–humus complex being one variant.

[18] The book by Hérody that I reference, purchased from D. Massenot after his presentation, focuses on viticultural planting but also elaborates on the fundamental agronomic principles of the Hérody method.

[19] Massenot and Hérody emphasize it is not merely the presence of organic matter that matters but rather its functionality, which varies with soil type and condition. In their view, fertility should not be confused with organic matter content. For instance, peat may contain 100% organic matter yet not be very fertile, while agricultural soil with only 1.7% organic matter can be highly fertile if microbial activity is robust.

there: among others, synergies between roots and fungi, as well as microbial support for plants. Massenot and Hérody have a rigid and myopic view of organic matter within agroforestry practices, which fundamentally embrace a dynamic understanding of ecosystem–soil relationships. Both agroforestry and agroecology aim to engage with natural regulatory processes, primarily by enhancing biodiversity in cultivated spaces. Here, organic matter is understood in terms of its role in the self-regeneration of soil fertility.

While Massenot and Hérody criticize conventional agriculture for its detrimental effects on soils, they also argue against agroecology—and particularly agroforestry—for elevating non-tillage to doctrinal status and for not adopting a pragmatic agronomic approach that could ensure adequate production. Although agroforestry typically opposes tillage, agroecology more specifically rejects its routine application. Furthermore, as we will discuss later, it is entirely possible to conduct economically sustainable "organic" viticulture without significant tillage, provided the terroir's geophysical characteristics allow it.

13.4 A "Naturalistic" Influence on Soil Management

For an anthropologist who isn't a specialist in agro-pedology, what is interesting about Massenot and Hérody's approach is that it is based on the dichotomy between "cultivated soil" and "uncultivated soil" (e.g., forest soil or fallow land):

> It is crucial not to conflate meadow-type organic accumulation with mulch-type organic accumulation; they operate differently, support distinct nutritional selections in plant populations, and certainly differ in their organo-mineral configurations. "Managed" grasslands, especially those with legumes,[20] provide a more efficient conduit for nitrogen than fallow land. Similarly, "sown" biodiversity is more beneficial for cultivated soils than spontaneous biodiversity. (...) How many times have I heard the "naturalizing" argument of spontaneous flora... (Hérody: 26, 36, 44)

The relevance of the term "spontaneous flora" warrants scrutiny, as it ostensibly exists only in the context of human activity. In areas untouched

[20] Legumes are referred to as green manure.

by human influence over extended periods, what develops is a diversified habitat characterized by complex interrelations. Massenot and Hérody advocate for an exploitative approach to living organisms, aiming to "optimize," "sustain," and be "efficient." Their analysis of microbial processes predominantly addresses physicochemical aspects confined to soil fertility within a productive framework, contrasting sharply with agroecological practices, which emphasize ecosystemic interdependencies[21] (i.e., the correlation between the physicochemical and the biochemical): interdependencies of microbes among themselves (e.g., bacteria, fungi, etc.) and with microflora and microfauna underscore the complexity of soil ecosystems. "The soil is alive; indeed, it is an ecosystem made up of the biodiversity of the organisms residing within it" (Selosse: 116). These soil experts appear to underestimate the significance of living organisms in the development of soil. Soil formation, as we know it today, began approximately 400 million years ago when plants developed roots, which allowed vegetation to proliferate exponentially. This growth led to an enhanced and more complex microbial life, subsequently accelerating the alteration of the bedrock (Selosse: 262).

The notion of "a soil that works" reflects an anthropocentric framing of our interaction with the living environment, in which humans oversee and manipulate soil processes, leading to the "objectification" of the soil and the non-human world at large. Avoiding the dichotomy of "nature" *vs.* "culture" that is inherent in the modern Western conception of the world, which Descola (2005) describes as "naturalistic," the debate over tillage revisits this dilemma, framed as non-intervention (nature) *vs.* intervention (culture). Descola emphasizes that "naturalism" refers to the relationship modern Westerners have with the world, where "nature" comprises a distinct set of phenomena and entities separate from human activities or "culture." He does not define "naturalism" as a pure and simple separation of nature and culture but strongly insists on the "intellectual and social consequences that the concept of nature has had [consequences] observable in practical life and in the forms of access to nature that have marked the industrial world" (Descola, 2014: 298).

As Hérody observes, "The cultivated vine is inseparable from the human endeavor aimed at producing a fruit for purposes not originally intended by Nature" (10). For Massenot and Hérody, cultivation involves

[21] In the early 1990s, when DNA research dramatically advanced our understanding of soil microbiology, 99% of microbial biodiversity was unknown. *Cf.* Selosse, p. 100.

interacting with and domesticating nature (here, the soil); hence, tillage aligns with cultural activity, while non-tillage aligns with a pursuit of an idealized natural state: "letting nature take its course." Thus, the act of cultivation inherently involves domesticating the living environment and managing living entities.[22]

For Massenot and Hérody, the soil is treated as a geochemical entity under human control, a resource to be actively managed. Thus, good soil is soil that works—specifically, soil that sustains fertility for perennial crops expected to meet certain productive yields. The goal is to "make the soil work at the optimum level permitted by its genetic potential" (Hérody: 25). Here, soil is not considered merely in terms of its functional capacity but constrained to a singular purpose, despite its multifaceted roles: It is "a support for living beings, a highly biodiverse habitat, a reservoir of organic and mineral substances, a regulator of ecosystem exchanges and flows, a site for the transformation of mineral and organic substances, and a purifier of toxins."

The idea of "letting nature take its course" is advocated by those who favor reducing or even abandoning tillage in some cases. This approach draws on what Haudricourt (1962) highlighted with the distinction between "positive direct action" and "negative indirect action." In the Melanesian context he studied, there is a respectful and "friendly" relationship with tropical tubers, which are treated individually rather than with harsh, uniform, and ongoing intervention. The term "indirect" does not imply neglect, such as allowing the spread of undesirable vegetation, or necessarily even less work, but rather an approach that focuses less on manipulating the plant directly and more on enhancing its surrounding environment to foster the optimal conditions for it to flourish. This perspective reflects a different view of human interaction with our living environment.

Finally, it is important to highlight that the agroecological design of viticulture (with agroforestry as one approach among others) perceives soil as a vast and intricate network of living entities interconnected by relationships of interdependence and mutual influence. This perspective encourages us to remain open to "the dizzying otherness of beings, the list of which is ever-expanding, and to the multiple ways in which they exist or relate to one another, without lumping them together too

[22] This influence is linked to the economic implications of cultivation. Hérody also discusses the concept of the "economic limit of viticultural unproductivity" (29).

quickly into any single category—and certainly not under the umbrella of 'nature'" (Latour, 2015: 50).

13.5 Minimizing Tillage and Engaging with Living Organisms Through Cover Crops

The adoption of cover crops in organic viticulture has expanded significantly over the past decade. These crops are used in various ways. Typically, they are integrated into every (or every other) inter-row, with tillage still occurring at the row level. The traditional use of "steel," or mechanical tillage, is thus partly supplanted by "vegetation," which serves multiple functions. In summary, the aerial parts of the plants, alongside the roots and fungi (the rhizosphere), primarily reduce soil erosion. Cover crops also enrich the subsoil with organic matter, nourishing the soil's microbial life. They also help maintain soil fertility[23] and, when legumes are planted, enhance nitrogen fixation, which benefits "the endomycorrhizal fungi associated with the vine roots. By forming a mycorrhizal network, these fungi can even transfer nitrogen from the legumes directly to the vine roots" (Selosse: 426). In addition, cover crops contribute to the structure of soil by improving its bearing capacity. They also provide an effective means of weed control.

Cover cropping requires minimal tillage and can be achieved through "direct" seeding—that is, by simply creating a furrow and broadcasting the seeds. However, they do induce a slight decrease in yield (usually less than 10%) and compete with vines for water access, though this is mitigated by enhanced soil moisture retention. Typically, cover crops are mowed in the spring and returned to the soil in a rough, partially buried manner.

[23] "Microbes responsible for mineralizing organic matter continually produce mineral salts: nitrates and phosphates. If not absorbed by roots or mycorrhizal fungi, they are carried away by water. Plant covers conserve this fertility by using it, then reincorporating it into the soil when they die." (Selosse: 422).

I would like to highlight the example of Sébastien David,[24] a winemaker based in Saint-Nicolas de Bourgueil, who shows that an interventionist approach to soil management, emphasizing the use of permanent cover crops and minimizing tillage, is viable in "organic" viticulture.

> My method is influenced by a variety of practices: biodynamics, permaculture, agroforestry, and the techniques developed by the Burgundians… My experiences are also shaped by extensive travel and reading, as well as my winemaking activities in Portugal, Georgia, and Armenia. (…) Since 2012, I've stopped traditional tillage: I don't plow or turn the soil. Instead, I make incisions of 3 to 4 centimeters deep with a serrated straight cutter. This can't strictly be called scratching; it's more about introducing slight aeration to the soil. (…) In the inter-row spaces, I don't turn over the grass but establish cover crops through direct seeding and select seeds from Eastern Europe tailored to the specific terroir (Fig. 1). Each variety of seed plays a distinct role and penetrates to different soil depths. My cover crops are perennial, and I perform late mowing when they are going to seed, thus avoiding the typical burial process. Initially, in 2015–2016, this approach was a bit hard on the vines, as the cover crops significantly utilized available resources. However, the primary advantage of this strategy lies in its holistic consideration of the ecosystem. The objective is to enhance biodiversity. I place great importance on the historical resilience of the vines, which has enabled me to reduce copper usage significantly. (…) Regarding the vegetation around the vine bases, we perform light hilling and dehilling once a year and use a vine weeder once or twice a year. (…) We also finely grind the vine prunings and introduce biodynamic compost. Despite these efforts, there is still a deficit of 15 units of nitrogen in the soil. It took seven years for the system to stabilize fully. I'm a firm believer in natural balances and often draw parallels with the forest floor ecosystem.

The philosophy behind permanent plant cover predominantly revolves around the concept of "negative indirect action." Direct seeding requires sporadic, light, and superficial tillage, but this is driven by a different rationale: It serves indirectly, merely to facilitate seeding. It is also noteworthy that Sébastien David does not completely forgo tillage; he restricts it to

[24] Sébastien David's estate has 15 hectares of Cabernet Franc, a varietal that defines the appellation. Coming from a family of winemakers dating back 15 generations in Saint-Nicolas de Bourgueil, David initiated the Syndicat des Vins Naturels and its "Vin Méthode Nature" label.

select, minimally intrusive actions. What stands out about this winemaker's approach is his principle of working with the ecosystem rather than against it; he adopts an interventionist approach that is attuned to the ecosystem's complexity, and his holistic view of the ecosystem heightens his sensitivity to the adverse and toxic effects of copper sulfate on living organisms. Thus, David opts to use copper sparingly, despite its prevalent use in organic viticulture as an antifungal agent to prevent fungal diseases, particularly because it significantly disrupts soil microbial life, including mycorrhizae. Overall, his commitment to agroecological practices means accepting a slight reduction in yield, which, however, does not threaten the economic viability of his estate.

Hence, respecting life does not mean doing nothing. When it comes to obtaining a harvest, the dichotomy between intervention and non-intervention is of little relevance. Why is it generally deemed undesirable to simply do nothing? I will address this question in part through my own experiences. In 2019, I purchased a 0.5-hectare vineyard that had been neglected for several years, located near Sainte-Maure de Touraine at the edge of the Chinon region. My initial goal in taking up viticulture was to reduce tillage and thereby foster a more vibrant soil ecosystem. Remaining true to my convictions, I restricted my activities to light manual hoeing at the level of the vine mound left "flat" (without any stripping or use of inter-row blades) and subsequently limited myself to clearing the undergrowth between the vine, after which I simply cleared the undergrowth between the vines. I mowed regularly in the inter-rows. The result was a plot resplendent with lush greenery and wildflowers, showcasing notable biodiversity, particularly among the microfauna. As a beginner, however, I encountered a significant challenge with couch grass, which thrived in the sandy soil of the parcel, an ideal environment for it. The soil was lively and well-structured, teeming with earthworms that thrived on the couch grass. Unfortunately, the vines were weak and produced very low yields, although the grapes themselves were of excellent quality. The couch grass was taking up all the nitrogen and other nutrients, as well as water, and its rhizomes were choking the vine roots. Therefore, in this context, the presence of "spontaneous" flora proved to be incompatible with sustainable vine cultivation. In response, in 2022, I started sowing specific cover crops, particularly grasses like rye in winter and buckwheat in spring, which competed effectively against the couch grass thanks to their extensive root systems. This approach is crucial as any tillage tends to stimulate the germination of weeds, a fact particularly true for rhizomatic weeds like couch grass, which are notoriously difficult to eradicate. Finally, given that sandy soils typically lack substantial organic material, the introduction of cover crops also helps enrich and structure the soil.

13.6 Conclusion

Contrary to popular belief, mechanical tillage does not inherently enhance soil life; rather, it primarily stimulates underground microbial activity (without polluting the water table), which is only one very specific aspect of the soil as a living environment. Thus, the widely assumed correlation between tillage and soil life is not straightforward and may even appear contradictory. This agricultural practice is deeply influenced by a "naturalistic" and overly simplistic view of the living world. The propagation of tillage practices, especially in educational settings, likely remains overly rigid and rooted in an outdated agronomic model that predates the emergence of contemporary environmental challenges while also being excessively reliant on the economics of viticultural machinery.

The agroecological approach, which advocates for minimizing tillage—particularly through the use of cover crops—demonstrates that it is feasible to develop an "organic" viticulture that prioritizes biodiversity without centering mechanical tillage. Beyond critiquing the impacts of tillage on soil life, agroecology fundamentally encourages organic winemakers to transform their relationship with the soil by adopting alternative methods and considering various approaches to cultivating economically sustainable viticulture that engages with the richness and complexity of life. This involves moving beyond the typical "naturalist" dichotomy of intervention *vs.* non-intervention, often synonymous with tillage *vs.* non-tillage. In agroecology, it is not so much the potentially minimal nature of intervention at the soil level that is interesting but rather the fact of intervening differently (which can require a lot of time and effort), through other methods, in order to establish a relationship of otherness with the living world.

Competing Interests and Acknowledgments The author would like to thank Céline Granjou for her kind and relevant comments, as well as the entire editorial team. Also thank you to INRAE which, thanks to Céline Granjou, agreed to finance the translation of this chapter.

Bibliography

Charbonnier, P. (2022). *Culture écologique*, SciencesPo (les presses).

Descola, P. (2005). *Par-delà nature et culture*, Gallimard (Bibliothèque des sciences humaines), Paris.

Descola, P. (2014). *La composition des mondes*, (entretien avec P. Charbonnier), Champs essais Flammarion.

Feller, C., Marsily, G., Mougin, C., Pérès, G., Poss, R., & Winiarski, T. (2016). *Le sol, une merveille sous nos pieds*, Belin (Pour la science), Paris.

Fournil, J., Kon Kam King, J., Granjou, C., & Cécillon, L. (2018). "Le sol : enquête sur les mécanismes de (non) émergence d'un problème public environnemental," *VertigO - la revue électronique en sciences de l'environnement* [online], vol. 18 no. 2.

Gobat, J. M., Aragno, M., & Matthey, W. (2010). *Le sol vivant*. Bases de pédologie—Biologie des sols. 3rd ed., revised and augmented. Coll. Gérer l'environnement. Les Presses polytechniques et universitaires romandes, Lausanne.

Goulet, F. (2008). *L'innovation par retrait : reconfiguration des collectifs sociotechniques et de la nature dans le développement de techniques culturales sans labour*, (PhD thesis), Université Pierre Mendès-France (Grenoble).

Haudricourt, A. G. (1962). "Domestication des animaux, culture des plantes et traitement d'autrui," in l'Homme, vol. 2, no. 1, pp. 40–50.

Hérody, Y. (2016). *Replanter une vigne*, BRDA ed. (Collection: "Les fondamentaux de l'agriculture").

Latour, B. (2015). *Face à Gaïa. Huit conférences sur le nouveau régime climatique*, La découverte/Les empêcheurs de tourner en rond.

Le Vin ligérien. (2012). Le trimestriel des vins de Loire, no. 1, April-May-June.

Natoli, J. (2013). *Guide pratique du vin bio. Réussir sa conversion en bio*, Dunod (La Vigne), Paris.

Selosse, M. A. (2021). *L'origine du monde. Une histoire naturelle du sol à l'intention de ceux qui le piétinent*, Actes Sud, (illustration: Arnaud, Rafaelian).

CHAPTER 14

Earthly Rhythms, Loops, and Leaks: Lessons from a Collective Investigation

Olivier Labussière, Germain Meulemans, Céline Granjou, Adrien Baysse-Lainé, and Pierre-Olivier Garcia

This book invites the reader to think with and from contemporary earthly troubles. The processes explored in the various chapters cut across conventional categories such as "land," "soil," and "underground." Engaging in a critical examination of these developments required a collective investigation open to various scientific communities, drawing inspiration from geosocial approaches to the underground, soil new materialisms, and planetary relational approaches. Our starting questions

O. Labussière (✉) · A. Baysse-Lainé · P.-O. Garcia
Pacte Social Sciences Research Center, University Grenoble Alpes, CNRS, Science-Po Grenoble, Grenoble, France
e-mail: olivier.labussiere@umrpacte.fr

G. Meulemans
CNRS, Centre Alexandre-Koyré, Paris, France

C. Granjou
National Research Institute for Agriculture, Food and Environment, Grenoble, France

© The Author(s), under exclusive license to Springer Nature Switzerland AG 2025

O. Labussière et al. (eds.), *Back to the Ground*,
https://doi.org/10.1007/978-3-031-88888-5_14

were: What does it mean to live on a de-stratified Earth? How can we understand re-stratification efforts, whether driven by citizen initiatives, large industries, or state-sponsored schemes? What kinds of practices and sensitivities to the Earth emerge from these experiences?

These questions were all posed, albeit in slightly different ways, by scholars working within the soil humanities (Krzywoszynska & Marchesi, 2020; Salazar et al., 2020), and by those working on geosocial formations (Clark & Yusoff, 2017) and political geology (Bobbette & Donovan, 2021). Yet, those working on soil and those working on the underground tended to do so in separate streams. In this book (and in the conference and workshop that led to it), we therefore sought to bring them together through case studies that transcend or complicate the horizontal distinctions usually made between soil and underground.

We conclude this collective inquiry by highlighting how the chapters contribute to reimagining strata. Rather than viewing strata as a vertical succession of geosocial layers, they are reimagined as the multiple, transversal circulations of matter and practices operating in the cycles and loops of water, carbon, and other material across the Earth's surface. In so doing, we aim to experiment with a shared vocabulary and framework that would make it possible to think of land, soil, and the underground together. From the exploitation of lithium in the Andean salt flats to the remediation of arsenic-contaminated mines and rivers in Southern France, the pioneering experiments in hot dry geothermal energy in Los Alamos to the development of composting businesses that support organic soil fertilization in the Paris region—these processes all reflect initiatives for soil- and underground-based ecological transitions. However, these processes do not engage with the Earth in the same way, especially when considering issues of knowledge and power, environmental justice, and risk. How do these various green promises differ from each other? This postface seeks to unpack these differences.

All contributions in this book engage, in different ways, with opening up investigations into processes of de-/re-stratification. What lies beneath is not solely underground, as Clark and Szerszynski (2021) have already shown. Dispersing into the atmosphere, spreading across rivers, and silently creeping into our bodies, the ground constitutes us, and we live within it. Expressions like "political geology" or "soil politics" capture only part of the chain of consequences activated by the multiple existences and agencies of the ground. Media studies scholar Citton (2017) proposed the notion of "politics of the ground" to approach our ecological background. Inspired by Paul Ricœur's distinction between the human *figure* and the *settings* from which it detaches itself, this notion emphasizes that while human stories usually focus on *figures*, a decentered point of view invites us to other kinds of companionship with what lies behind and below. Thinking downward requires less a sense of verticality than an attentiveness and sensitivity more attuned to the ecological back- and below-ground. It involves (re)grounding our ways of perceiving and thinking about the world, as well as doing politics. In this sense, our book's title—*Back to the ground*—is less about moving back down to the Earth and more about paying closer attention to the political, often precarious, entanglements through which people and the ground connect and co-constitute one another.

Following the intersecting circulations of biogeochemical cycles—of water, carbon, nutrients, and so on—allows us to observe how knowledge and power evolve in relation to the uncertain transformations of soils and undergrounds. Our project builds on the work of Yusoff (2017) and Clarke and Yusoff (2017), who redefined geopolitics as the sociopolitical capitalization of the potentiality of matter across strata, rather than focusing solely on territory and scale. The case studies presented here contribute to understanding the permeability between strata, the mutability of matter, and the vitality of living organisms. The processing of water, carbon, or nutrients from bottom to top and vice versa creates experimental conditions in which many entities become active and produce unanticipated effects—thus troubling our taken-for-granted categories (e.g., animate/inanimate, solid/fluid, mineral/vegetal, etc.) (Ingold & Simonetti, 2021). It also draws attention to what is left after extraction. Beyond and following extractive processes, residues circulate, agglutinate, enter new cycles, and travel through space and time (Ureta & Flores, 2022). From this perspective, new "politics of the ground" emerge and invite us to adapt our thinking further to capture our shifting relationships with our ecological back- and below-grounds.

Throughout the contributions, the notions of "loops," "leaks," and "rhythms" emerge as critical tools for analyzing how and why the processes of making strata are political. This can be preliminarily illustrated by the exploitation of lithium in the Andean salt pans—specifically, the development of intensive "loops" (i.e., pumping and reinjecting water underground) driven by the boom in electric mobility. These loops disrupt the water "cycle" and give rise to industrial evaporation landscapes designed to concentrate lithium. The concepts of "loops," "leaks," and "rhythms" materialize our relationships with broader earthly "cycles" in space and time (e.g., water cycle, carbon cycle, etc.). From a relational point of view, the "loops" we generate in the pursuit of ecological transition represent a process of *entre-capture* in the sense that this process makes earthly "cycles" part of the infrastructure of our ways of life while also embedding ourselves in an unpredictable material world. In the industrial age, what we didn't have to worry about was infrastructure, as the consumer only had a relationship with the final good or service. By contrast, in the post-industrial and Anthropocene eras, infrastructure becomes an experimental ecological condition, open to planetary interactions. The notion of "rhythm" adds depth to this analysis by extending the study of geo-socio-technical systems to include physical realities (e.g., seasonal alternation), biological realities (e.g., rest, growth), and social realities (e.g., work, leisure) that interact with these systems (Walker, 2021; Labussière et al., 2025).

Below, we present the lessons learned from the chapters in this book regarding the initiatives and challenges faced during these times of earthly troubles.

14.1 Inhabiting a De-stratified Earth

Several of the book's chapters contribute to the debate on the habitability of the Earth (Gaillardet, 2023), not through the prism of global environmental changes and their impacts but through the lens of de-stratification. What does it mean to inhabit a de-stratified Earth? Literature, painting, photography, and the social sciences provide complementary points of view.

Gaëlle Delétraz, Mathilde Joncheray, and Delphine Montagne (Chapter 4) show us that in science fiction, living in a de-stratified environment often means being confined underground, far from the damaged world on the surface (Chapter 4). This genre aligns with modernist approaches to adaptation, described as a politics of immunity (Garcia, 2015)—in the sense of making ourselves disconnected and unconcerned by the ecological consequences of our actions. These imagined worlds strongly echo contemporary "burying practices," which involve disposing of waste underground, effectively transferring the impacts of human activities to remote spaces and times (Kearnes & Rickards, 2017).

Exploring new ways of circulating between strata is central to inhabiting a de-stratified world and moving beyond notions of vertical hierarchy, as Julie Beauté shows (Chapter 5). Yarka's paintings challenge "the basic kinesthetic experience of having our feet planted on the earth" (Clark, 2011, p. 6). They unsettle the categories that help us orient ourselves in our surroundings, including directing one's gaze downward or upward and perceiving matter and bodies (e.g., far/near, solid/friable, fast/slow). When ordered around a perspective, traditional painting brings all objects into a shared scale, offering them a common world in which to appear. Yarka, however, does not presuppose such a common world. His paintings invite us to imagine esthetic codes for a different, diffracted world, where unusual continuities and circulations unfold between subsoil, soil, and atmosphere. Echoing the notions of loops and leaks, Yarka's work prefigures worlds where strata are defined by unexpected interactions (e.g., disintegration, overflow, or permeability) that often reveal animacy (e.g., terrestrial activity or animality) within what might otherwise be perceived as inert.

For Olivier de Sépibus Thomaïdis (Chapter 3), recreating mediations becomes essential to learning how to adapt to a shifting ecological background in a de-stratified context (Chapter 3). Distanced from the iconic and now obsolete "figure" of the White Mountain, the melting glaciers are generating a new ground composed of vast areas of scree and remnants of moraines, which remain largely unnoticed. This experimentation with new ecological mediations takes place during a period of earthly troubles. It represents a critical step in a long-term process that aims at revising our relationship with the mountain—both culturally and legally—to foster greater attention and protection. However, this effort faces cultural resistance, as most art centers and media outlets favor outdated images of what the mountain is, perpetuating a lack of interest in de-stratified landscapes.

The de-stratified experience highlights the need for revisiting expert cultures and methodologies to better understand and study the ground (Chapter 6). Christelle Gramaglia and Sijia Du (Chapter 6) illustrate this through a case of conflicting methodologies in assessing arsenic-polluted soils. Expanding the assessment beyond the mining space—traditionally viewed as an independent extractive process from bottom to surface—to include broader ecological cycles shifts attention toward the importance of tracking material loops and open-ended flows of widespread pollution. Rather than a purely scientific challenge, this reframing provides a new scale to measure the impacts and responsibilities associated with former mining activities. It also challenges older expert cultures of risk management, which tend to focus on zoning and the containment of pollutants.

14.2 Earthly Re-stratifications

One of the forces that most visibly cut across Earth's layers is the water cycle, which permeates the atmosphere, soils, and the underground. In addition to forming its own compartment within the ecosystem (the hydrosphere), water serves as a dynamic link between other compartments (atmosphere, biosphere, lithosphere, pedosphere). Water continuously participates in the making and remaking of strata as it is transformed during its journey through them. The stories assembled in this book shed light on patchy sites of water circulation and stasis—rather than a unified water cycle—that intertwine (and are made to intertwine) with human histories. While bodies of water, such as rivers and seas, are often viewed as "natural" borders in horizontal territorial thinking, paying attention to water circulation across strata encourages consideration of the vertical and voluminous, downward extensions of modern territorial governance (Billé & Battaglia, 2020; Elden, 2013).

The three chapters explore different sites where tensions between the water cycle and the development of sociotechnical loops involving the underground can be observed. The water cycle is successively approached through the lenses of infiltration into urban soils, heat recovery, and lithium extraction. These processes occur in distinct geographical and political contexts; nevertheless, their combined analysis sheds light on the porosity of soils and the underground. Porosity challenges the various forms of expertise and knowledge that are typically compartmentalized and associated with specific layers. It enables transverse explorations in which soil, underground, and atmosphere may influence one another.

The first project of re-stratification, described by Maxime Algis (Chapter 7), is set in the City of Paris (France) (Chapter 7), where public policies aimed at reducing soil sealing and increasing rainwater infiltration challenge the sectorial management of urban soils. This involves restoring the water cycle through a stack of various sociotechnical systems from different eras: surface (roads), subsurface (supply water and energy networks), and underground (subway, car parks). While the Roads Department fears a return to a pre-hygienic environment (characterized by poorly drained rainwater and muddy urban spaces), the subway service is concerned about a higher risk of flooding—a restored water cycle is expected to avoid unwanted leaks. Translating the natural water cycle into successive adaptations is necessary to make water circulation compatible with different sociotechnical strata of the urban environment. However, this translation effort faces challenges from the sectorization of expertise, corporatism, and sociotechnical infrastructures.

The following two contributions delve into contemporary reconfigurations of extractivism and examine how various actors constitute strata, circulations, and thresholds between them to integrate these into the market economy. They shed light on how the multidirectional rhythms and cycles of the geophysical sphere are infrastructurally aligned with the rhythms of global capitalism and how earthly loops and rhythms are brought into the temporality of capitalism (and vice versa). As Yusoff observes, "capitalism succeeds as a geosocial machine because it organizes modes of capture that capitalize on geopower" (2017, p. 113). These contributions demonstrate how this co-dependency persists even after the resource is depleted or rendered obsolete by market transformations. When extraction ends, strata do not "return to nature." Instead, they remain active at intensities that are often unpredictable and require constant surveillance. As such, they remain a political issue, which becomes particularly salient when new forms of capitalization redefine their potential and new rounds of infrastructuring emerge—often following patterns of path dependency rooted in older knowledge and infrastructures that preceded them in the same places.

In Chapter 8, Alain Nadaï, Julien Merlin, and Olivier Labussière follows the development of a geothermal concept based on the creation of a man-made reservoir in bedrock, from its early experiments in Los Alamos in the 1970s to industrial development in France during the 1990s. A geothermal stratum was initially conceived as the result of creating a "loop" of water (injected at one point and pumped at another) in bedrock understood as a watertight and homogeneous environment. In this model, the bedrock is passive, while the water circulation actively forms the "loop." This paradigm was called into question by further developments in Soultz-sous-Forêt (France), where the bedrock proved to be faulted. A colored tracer revealed that the water pumped out differed from the water injected, showing that the geothermal "loop" was not closed but instead opened to "leaks" and mysterious circulations underground. The situation challenges geoengineers' scientific assumptions about the bedrock. In this case, earthly re-stratifications occurred despite an incomplete understanding of the geological milieu and its hydrogeology. The use of hydraulic fracturing to develop deep geothermal "loops" (Enhanced Geothermal System) has proved to be risky and harmful in certain geological environments.

For their part, Marie Forget, Vincent Bos, and Chloé Artero-Nicolas (Chapter 9) centers on the exploitation of lithium in the Andean salt pans. Owing to its legal classification, brine is managed differently from water, which allows the industry to create a geo-socio-technical "loop" by pumping the water table, using solar evaporation to concentrate the lithium salts, injecting seawater transported from the coast underground, and repeating the process. By being decoupled from the natural water cycle, this industrial loop is made scalable. It allows the industry to accelerate the production of lithium beyond the aquifer's regeneration capacity, which creates water imbalances that conflict with other social uses and ecological needs across the Andes. This critical case illustrates how the making of a stratum re-assembles the underground, soil, and atmosphere, shedding light on emerging political issues throughout the loop. The distinction between "cycle" and "loop" is not a dualistic partitioning between nature and culture but instead illuminates blurred geo-ontologies.

14.3 Emerging Practices and Sensitivities to the Earth

Shifting away from the excesses of mining capitalism and the management of its remnants, the contributions in the third part of the book invite us to explore another set of projects that aim to "re-loop" the cycles intertwining soils and agricultural activities. By engaging in discussions within the soil humanities (de la Bellacasa, 2017; Krzywoszynska, 2020), this section examines debates about how to care for the soil and soil life. These contributions focus on the reconfiguration of ways of knowing and describing soil in technical terms—whether derived from science and engineering or other forms of practice. They illustrate, in Yusoff's words, how "what is sayable and seeable is contingent on the mode and milieu of speech that organizes the stratum" (2017, p. 116).

These contributions show the trajectory and uses of categories such as "soil metabolism" or "soil life," which complicate a common narrative that tends to present these notions as necessarily antagonistic to the purely utilitarian perspectives on soil that predominated in soil science and conventional agriculture during the twentieth century. Vital perspectives on soil, while largely supported by anti-productivist currents, are now being incorporated into various ecological intensification projects. The living beings in the soil are increasingly regarded as a workforce and a tool for development and planning. Soil life and metabolism lie at the core of new forms of engineering storytelling that seek legitimacy for contemporary "re-looping" projects by imagining continuities with ancient practices of fertilization or material reuse—practices that, however, differ significantly in terms of scale and objectives.

Céline Pessis (Chapter 10) traces the trajectory and transformations of the notion of "living soil" in twentieth-century French agriculture. The "living" character of soil—inhabited by a host of fungi, insects, worms, and microbes—became largely invisible in post-war soil science and soil management methods, despite its emphasis in most academic circles during the first half of the century. In the post–World War II era, this "concern for the soil" mainly persisted in alternative agricultural practices, framed as a response to the neglect of soil life during the large-scale agricultural modernization movement in France, which relied heavily on pesticides and synthetic fertilizers. The chapter shows how controversies between organic and industrial agriculture revolve around different conceptions of what constitutes "a soil," leading to radically different approaches to making it (over)productive and involving different kinds of attention, promises, and sociotechnical projects.

Jérôme Gidoin (Chapter 13) examines how re-stratification is experienced in the world of professional viticulture through evolving practices of care toward soil life. He describes two competing forms of soil care in alternative viticulture: One uses mechanized plowing to prepare the soil while minimizing chemical inputs, and the other seeks to reduce plowing in order to better engage with natural soil-regulating processes and matter cycles. Breaking with conventional agronomic knowledge and taking charge of soil life transforms the ecological background—the wine-growing ecosystem as a whole—into an object of care and attention. This approach re-establishes vibrant subterranean life in close interaction with epigenetic flora and fauna and the transversal biogeochemical cycles that sustain vine fertility.

The final two chapters of the book suggest another kind of re-stratification, conceived as a way to repair (or care for) the metabolic loops that run through bodies (human and animal), plants, and soils. The experiments in returning human excrement to the soil, described by Marine Legrand, Etienne Dufour, Mathilde Soyer, Alexsandro Arbarotti, and Marc Higgin (Chapter 11), show how activists and engineers collaborate to envision and implement human composting as a form of co-production between humans and soils. The treatment of human excreta in modern sanitation systems, which consume a great deal of water and degrade rivers through excessive nutrient influx, could instead direct these nutrients to nourish the soil. The projects they describe are infrastructural experiments (or experiments in de-infrastructuring) designed to integrate the human body into cycles perceived as more "natural." These experiments create a fertile geosocial stratum by modifying bodily practices, hygiene, and diet to revisit the question of what remains, its qualities, and its usefulness for other non-humans. Here, the relationship between strata and cycles comes to the fore again: improving compost requires the realignment of bodies (e.g., stopping drug use). The human body is seen as both a source of fertility and a risk of toxicity.

Finally, Maud Hetzel's chapter (Chapter 12) explores another effort to restore soil's metabolic loops, with a description of the institutionalization of composting practices and their integration into the market economy—practices that were previously confined to alternative or even subversive niches. In France, composting is now an emerging sector that is currently adopting industry-style norms and standards. This includes the establishment of a new target of "compost stabilization," where compost is designed to become a standardized, predictable, and interchangeable material. In this process, food waste and overproduction are redefined in terms of their economic potential. However, uncertainties remain regarding how best to maximize this resource, as collection and recycling circuits fall under the purview of different professions and compete with other sectors (such as methanization) for access. These efforts also raise questions about individual vs. collective outlets, which entail different stakes and values and ultimately reflect divergent approaches to transforming soil into a shared or unshared fertile stratum.

Following these analyses, ground-based promises of a green transition do not appear to be inherently positive or negative. The chapters shed light on the relational dynamics that arise from contemporary processes of making strata. For example, creating a geothermal "loop" into the bedrock leads to unexpected interactions with faults and deep-water tables, while (re)creating a metabolic "loop" connecting human and animal bodies to the soil brings about both the chance of renewing how we think about and perform "fertility," and new risks of toxicity. Thus, ground-based promises possess not only technical potential but also *relational* potential.

Focusing on relationships allows us to distinguish between projects that seek to align multiple loops and rhythms (e.g., revising our relationships with food, medicine, and the human body to create organic fertilizers) and those that overlook such interconnected concerns. This relational potential can become transformative when the systemic effects of an initiative are taken into account. By contrast, some projects merely exploit pre-existing terrestrial cycles to develop productive and scalable "solutions" (e.g., disrupting water cycles to intensify lithium production). From a relational point of view, the latter can be characterized by their disruptive potential. Etymologically, disruption refers to the breaking of links. A disruption is more than an isolated disturbance and less than a generalized collapse (Stiegler, 2019). It affects the interdependencies within socio-ecosystems. Environments lose aspects of their information, historicity, and uniqueness. As a result, they are less capable of maintaining equilibrium, which renders them more random and vulnerable. The Anthropocene is accelerating the entropic unraveling of the world. What is disrupted is not merely the product of a single history but the capacity to create new histories with the Earth.

Bibliography

Billé, F., & Battaglia, D. (2020). *Voluminous states: Sovereignty, materiality and the territorial imagination*. Duke University Press.

Bobbette, A., & Donovan, A. (2021). *Political geology*. Oxford University Press.

Citton, Y. (2017). *The ecology of attention*. Polity Press. (Original work published 2014).

Clark, N. (2011). *Inhuman nature: Sociable life on a dynamic planet*. Sage.

Clark, N., & Yusoff, K. (2017). Geosocial formations and the Anthropocene. *Theory, Culture & Society, 34*(2–3), 3–23.

Clark, N., & Szerszynski, B. (2021). *Planetary social thought: The Anthropocene challenge to the social sciences*. Polity Press.

de La Bellacasa, M. P. (2017). *Matters of care: Speculative ethics in more than human worlds*. University of Minnesota Press.

Elden, S. (2013). Secure the volume: Vertical geopolitics and the depth of power. *Political Geography, 34*, 35–51.

Gaillardet, J. (2023). *La terre habitable : ou l'épopée de la zone critique*. La Découverte.

Garcia, P. O. (2015). *Under adaptation, immunity: Study on the discourse of climate change adaptation*. [Doctoral dissertation, University of Grenoble Alpes].

Ingold, T., & Simonetti, C. (2021). Introducing solid fluids. *Theory, Culture & Society, 29*(3), 3–29.

Kearnes, M., & Rickards, L. (2017). Earthly graves for environmental futures: Techno-burial practices. *Futures, 92*, 48–58.

Krzywoszynska, A., & Marchesi, G. (2020). Toward a relational materiality of soils: Introduction. *Environmental Humanities, 12*(1), 190–204.

Krzywoszynska, A. (2020). Nonhuman labor and the making of resources: Making soils a resource through microbial labor. *Environmental Humanities, 12*(1), 227–249.

Labussière, O., Merlin, J., & Nadaï, A. (2025). Orchestrating deep geothermal energy policy? The exploitation of the Dogger aquifer in the Paris Basin (1950–1990). In P. Hamman & N. Roudil (Eds.), *The future of cities and energies in Western Europe: Scales, actors and social experiments*. De Gruyter.

Salazar, J. F., Granjou, C., Kearnes, M., Krzywoszynska, A., & Tironi, M. (2020). *Thinking with soils: Material politics and social theory*. Bloomsbury Publishing.

Stiegler, B. (2019). *The age of disruption: Technology and madness in computational capitalism*. Polity Press.

Ureta, S., & Flores, P. (2022). *Worlds of gray and green: Mineral extraction as ecological practice*. University of California Press.

Walker, G. (2021). *Energy and rhythm. Rhythmanalysis for a low carbon future*. Rowman & Littlefield.

Yusoff, K. (2017). Geosocial strata. *Theory, Culture & Society, 34*(2–3), 105–127.

Names of Authors

© The Editor(s) (if applicable) and The Author(s), under exclusive license to Springer Nature Switzerland AG 2025
O. Labussière et al. (eds.), *Back to the Ground*,
https://doi.org/10.1007/978-3-031-88888-5

Names of Relevant Countries, Regions, Jurisdictions

© The Editor(s) (if applicable) and The Author(s), under exclusive license to Springer Nature Switzerland AG 2025

O. Labussière et al. (eds.), *Back to the Ground*,
https://doi.org/10.1007/978-3-031-88888-5

Subject Index

© The Editor(s) (if applicable) and The Author(s), under exclusive license to Springer Nature Switzerland AG 2025
O. Labussière et al. (eds.), *Back to the Ground*,
https://doi.org/10.1007/978-3-031-88888-5

S

T

U

FSC
www.fsc.org
MIX
Papier | Fördert
gute Waldnutzung
FSC® C083411

Kolibri 360 GmbH
Ferdinand-Jühlke-Straße 7
99095 Erfurt, Deutschland
produktsicherheit@kolibri360.de